GEOGRAPHY AND ECONOMY IN SOUTH AFRICA AND ITS NEIGHBOURS

Geography and Economy in South Africa and its Neighbours

Edited by

ANTHONY LEMON
School of Geography and the Environment, Oxford University

CHRISTIAN M. ROGERSON
Geography Department, University of the Witwatersrand, South Africa

Routledge
Taylor & Francis Group

LONDON AND NEW YORK

First published 2002 by Ashgate Publishing

Published 2017 by Routledge
2 Park Square, Milton Park, Abingdon, Oxfordshire OX14 4RN
711 Third Avenue, New York, NY 10017, USA

First issued in papcrback 2017

Routledge is an imprint of the Taylor & Francis Group, an informa business

British Library Cataloguing in Publication Data
Geography and economy in South Africa and its neighbours. -
  (Urban and regional planning and development)
  1.South Africa - Economic conditions - 1991- 2. Africa,
  Southern - Economic conditions - 1994-
  I.Lemon, Anthony II.Rogerson, C. M. (Christian Myles)
  330.9'68'065

Library of Congress Cataloging-in-Publication Data
Geography and economy in South Africa and its neighbours / edited by Anthony Lemon and Christian M. Rogerson.
    p. cm. − (Urban and regional planning and development series)
  Includes bibliographical references and index.
  ISBN 0-7546-1868-4
    1. South Africa--Economic conditions--1991- 2. Africa, Southern--Economic conditions--Case studies. I. Lemon, Anthony. II. Rogerson, C. M. (Christian Myles) III. Urban and regional planning and development

HC905 .G467 2002
330.968--dc21   2002018636

ISBN 13: 978-1-138-25825-9 (pbk)
ISBN 13: 978-0-7546-1868-3 (hbk)

Contents

PART 1: SOUTH AFRICA

List of Figures

List of Tables

List of Contributors

Patrick Bond is a geographer and political economist educated at Johns Hopkins University who is Associate Professor in the Graduate School of Public and Development Management at the University of the Witwatersrand, Johannesburg. His books include *Unsustainable South Africa* (University of Natal and Africa World Press, 2002), *Zimbabwe's Plunge* (University of Natal, Weaver, Africa World Press and Merlin, 2002), *Against Global Apartheid* (University of Cape Town, 2001), *Elite Transition* (Pluto Press, 2001), *Cities of Gold, Townships of Coal* (Africa World Press), and *Uneven Zimbabwe* (Africa World Press, 1998). In addition to extensive policy drafting and advocacy, his work focuses on discourses associated with movements for economic, social and environmental justice, which he serves through the Municipal Services Project (www.queensu.ca/msp), the Alternative Information and Development Centre (www.aidc.org.za) and the Center for Economic Justice (www.worldbankboycott.org). He has held visiting professorships at York University (Canada) and Yokohama National University, was an assistant professor at Johns Hopkins University School of Public Health, and worked for several years in two Johannesburg NGOs.

Paul Crankshaw is the founder and publisher of *African Mining*, South Africa's leading business-to-business magazine for the mining industry. He is also managing director of Johannesburg-based Brooke Patrick Publications (Pty) Ltd, which publishes *African Mining* and other mining-, construction- and infrastructure-related journals. He holds a Bachelor of Journalism and Media Studies degree from Rhodes University and Honours and Masters degrees in Development Studies from the University of the Witwatersrand. He is currently based in the UK, where he runs Brooke Patrick Publications UK and is responsible for the company's business development in the UK and Europe.

Michael Bernard Kwesi Darkoh is Professor and Head of the Department of Environmental Science at the University of Botswana. He is an economic geographer whose research interests include environment and sustainable development, land degradation and desertification, industrial location and manufacturing in developing countries. He has published extensively and his books include *Tanzania's Growth Centres* (Peter Lang,

1994), *African River Basins and Dryland Crises* (Uppsala University and OSSREA, 1992) and *Combating Desertification in the Southern African Region* (UNEP, 1989). His most recent book is an edited reader on *Human Impact on Environment and Sustainable Development in Africa* (Ashgate, 2000).

Richard Gibb is Reader in Human Geography at the University of Plymouth, England. His research interests focus on regional economic and political integration, with a particular focus on southern Africa and the European Union. His latest research examines southern Africa's trading relationships with the EU under the Cotonou Agreement and the South Africa-EU Trade, Development and Co-operation Agreement. He is co-author (with Mark Wise) of *Single Market to Social Europe* (Longman, 1993) and co-editor (with W.Z. Michalek) of *Continental Trading Blocs: the Growth of Regionalism in the World Economy* (John Wiley, 1994).

Graham Harrison lectures in politics at the University of Sheffield. He has made repeated research visits to Mozambique since the mid-1990s, publishing in the *Review of African Political Economy*, *Democratization*, and the *Third World Quarterly*. He has also written a book on *The Politics of Democratisation in Rural Mozambique*. His other research interests include the World Bank, administrative reform in Uganda and Tanzania, and the political economy of development. He is editor of the *Review of African Political Economy* and *New Political Economy*.

Meshack Khosa holds a D. Phil. from Oxford University and an MA from the University of the Witwatersrand. He has taught urban studies at the University of Natal, Durban and was Global Security Research Fellow at the University of Cambridge in 1995. He subsequently worked for the Human Sciences Research Council (HSRC) in Pretoria, latterly as research director of the Democracy and Governance Group, and is now Director of the MTN Foundation. He has published extensively on aspects of transport policy, the taxi industry, land reform, health restructuring and regionalism in South Africa, and edited several books for the HSRC.

Anthony Lemon is University Lecturer in the School of Geography and Environment, Oxford University and a Fellow of Mansfield College. He has held visiting lectureships and research fellowships at the Universities of Natal, Rhodes, Zimbabwe, Cape Town, Witwatersrand and Stellenbosch. He has a longstanding research interest in the geography of apartheid and post-apartheid restructuring, on which he has published extensively. He is author of *Apartheid: a Geography of Separation* (Saxon House, 1976) and

Apartheid in Transition (Gower, 1987), and has edited *Homes Apart: South Africa's Segregated Cities* (Paul Chapman, David Philip and Indiana University Press, 1990) and *The Geography of Change in South Africa* (John Wiley, 1995). He also co-edited, with Norman Pollock, *Studies in Overseas Settlement and Population* (Longman, 1980). His current research interests centre on desegregation and redistribution in post-apartheid education.

Charles Mather is Senior Lecturer in Human Geography, School of Geography, Archaeology and Environmental Studies, University of the Witwatersrand, Johannesburg. His research interests focus on the restructuring of South African agriculture after apartheid and he has published several pieces on the relationship between agricultural restructuring and rural livelihoods and farm labour. He is currently involved in a detailed project of South African citrus exports after deregulation.

Miranda Miles-Mafafo is Lecturer in Human Geography, School of Geography, Archaeology and Environmental Studies, University of the Witwatersrand, Johannesburg. She has published widely on development issues in Swaziland. Her research interests include the socio-economic aspects of post-colonial female migration and Swazi women's livelihood strategies and roles in domestic work and housing.

Agnes Musyoki is Professor of Human Geography at the University of Venda for Science and Technology, South Africa, and has previously taught at universities in Kenya and Botswana. She holds a PhD degree from Howard University, USA and an MA degree from Ohio University. Her areas of research and publication include rural market studies, regional development planning, and gender and development.

Etienne Nel is Associate Professor of Geography at Rhodes University in Grahamstown, South Africa. He was born in Zambia and completed his university education at the Universities of Rhodes and the Witwatersrand. His doctoral thesis was subsequently published as *Regional and Local Development in South Africa: the Experience of the Eastern Cape* (Ashgate, 1999) and he has published many papers on South African economic geography, local economic development and community development.

Christian M. Rogerson is Professor of Human Geography, School of Geography, Archaeology and Environmental Studies, University of the

Witwatersrand, Johannesburg. Issues surrounding local economic development, tourism, small enterprise development and regional change represent his major research foci. He is the author of over 180 published articles on aspects of economic development in southern Africa and has co-edited (with Eleanor Preston-Whyte) *South Africa's Informal Economy* (Oxford University Press, Cape Town, 1991) and (with Jeffrey McCarthy) *Geography in a Changing South Africa* (Oxford University Press, Cape Town, 1992).

David Simon is Professor of Development Geography and Director of the Centre for Developing Areas Research at Royal Holloway, University of London. He has published widely on development theory and policy, post-colonial urban and regional change, transport and the environment-development interface, with particular reference to sub-Saharan Africa. He is also a longstanding Namibian specialist. His most recent books are (as editor and co-editor respectively) *South Africa in Southern Africa: Reconfiguring the Region* (James Currey, David Philip, and Ohio University Press), and (with Anders Närman) *Development as Theory and Practice: Current Perspectives on Development and Development Co-operation* (Addison Wesley Longman, 1999).

Lance van Sittert is a Senior Lecturer in the Department of Historical Studies at the University of Cape Town. He works in the field of environmental history with a particular emphasis on the marine environment.

Daniel Tevera is a Zimbabwean geographer with a doctorate from Cincinnati, USA. He is Associate Professor in the Department of Geography and Environmental Science at the University of Zimbabwe and has held visiting lectureships in Gothenburg, Botswana, and Keele, England. He has published extensively on urban issues and regional development problems in southern Africa and has co-edited (with S. Moyo) *Environmental Security in Southern Africa* (SAPES Books, Harare, 2000) and (with L. Zinyama and S.D. Cumming) *Harare: the Growth and Problems of the City* (University of Zimbabwe Publications, 1993).

Lovemore Zinyama taught in the Department of Geography and Environmental Science at the University of Zimbabwe in Harare from 1977 until 1999. He held the Chair of Geography from 1992 until he left to join the private sector in 1999. He was educated at the University of Zimbabwe where he obtained his doctorate in agricultural geography and at the School of Development Studies at the University of East Anglia, UK. He has published extensively on agriculture and rural development in Zimbabwe, urban development, and on internal and regional migration of population.

Acknowledgements

We should like to thank first and foremost all our contributors for their support, time and patience in answering many editorial questions. We owe a great debt to Jane Battersby of the School of Geography, Oxford University for assisting us to meet the requirements of camera-ready copy. We should also like to thank our cartographers, Ailsa Allen and David Sansom of the School of Geography, Oxford University (Figures 3.2, 3.3, 10.1, 11.1, 12.1, 13.1, 14.1, 16.1), Wendy Job, University of the Witwatersrand (Figures I.1, I.2), Brian Rogers, Plymouth University (Figures 16.2-16.5) and Jenny Kynaston, Royal Holloway, London (Figure 9.1).

Note on Currency

The dollar ($) refers to United States dollars throughout the book, unless otherwise specified.

Geography and Economy in South Africa and its Neighbours: An Introduction

ANTHONY LEMON AND CHRISTIAN M. ROGERSON

Introduction, Context and Objectives

This volume focuses on the changing economic geography of southern Africa, which is defined here as constituted by South Africa and its six neighbouring states of Botswana, Lesotho, Mozambique, Namibia, Swaziland and Zimbabwe (Figure I.1). Our study area of southern Africa represents one of the poorest regions in the world. Estimates made for 2000 suggest that approximately half of the region's total population are living on the proverbial US 'dollar a day or less'. Poverty is reflected in weak social indicators, such as high levels of malnutrition, illiteracy, unemployment, underemployment and declining life expectancy as well as unsatisfactory access to basic services and infrastructure (Ramsamy, 2001). The spread of the HIV/AIDS pandemic across the region has resulted in some of the world's highest recorded infection rates, further exacerbating the problem of poverty. For example, South Africa has one of the highest infection rates in the world with 4.7m. people (one in nine of the country's population) recognised as HIV positive, a level of infection that is expected to increase dramatically the numbers of households living in poverty (Lewis, 2001). Although the majority of the poor in Southern Africa reside in rural areas, the wealthier cities and towns of the region show striking signs of worsening conditions of urban poverty. Whether in Maseru, Harare, Maputo, Mbabane or Johannesburg, there are to be observed common symptoms of economic distress in the form of visible unemployment, growing levels of crime and mushrooming zones of informal shelter which often are devoid of basic services of water or sanitation.

It is against this overarching context of poverty that there have been a number of high profile statements and initiatives made by African political

Figure I.1 South Africa and its Neighbours: Key Location Features

leaders to address the region's economic problems and to foster a new climate for renewal and revival. Foremost amongst these new ideas are those that surround the 'mythical' claims for an 'African Renaissance' (Olivier, 2001). At its root it is hoped that the liberation of South Africa from apartheid presents an opportunity for Africa to break with its past condition of underdevelopment. The core objective of the African Renaissance is to get rid of poverty and misery, empowering the masses economically and socially through participatory democracy and good governance. In its most recent variant, the concept of the African Renaissance has received more concrete expression in what is known as the 'New Partnership for Africa's Development' (NEPAD), previously known as the New Africa Initiative. The central ideas behind this particular initiative, which is strongly endorsed by South African President Thabo Mbeki, concern the eradication of poverty and the placing of Africa on a sustainable trajectory of development (Olivier, 2001; Spicer, 2001). Good governance and linkages among Africa's states and leaders are essential ingredients for success.

Immense challenges face the seven countries covered in this book in attaining these worthy objectives of the African Renaissance. At the broadest level, the economies of the seven countries share several common characteristics in terms of systematic underdevelopment, poverty and socio-economic inequality. Most importantly, in all countries the most dynamic and efficient part of the economy, which accounts for the bulk of regional Gross Domestic Product, is the formal sector. Nevertheless, with the exception of South Africa, this sector provides jobs for only one-fifth of the region's labour force (Lambrechts, 2001), and this share is tending to decrease rather than increase amidst the economic climate of the early 2000s. Accordingly, the informal economy represents the major source of incomes and livelihoods for a large segment of the region's population in both rural and urban areas (Rogerson, 1997). The United Nations Economic Commission for Africa asserts that the pace, consistency and level of economic growth taking place in the region continue to lag behind that of the growth of population, estimated as an annual 3.5 per cent. Indeed, it estimates that a GDP growth of 6.2 per cent with a level of investment of up to 38 per cent of GDP, is needed to reduce poverty by half by a target year of 2015 (Ramsamy, 2001). Set against these growth targets, the economic performance of the seven countries for the period 1991-98 has been disappointing; the best growth performers have been Mozambique (7.6 per cent per annum) and Botswana (5 per cent). For the period 1991-98, the weakest economic records have been posted by South Africa, Zimbabwe and Swaziland (ibid.). Indeed, at the turn of the new millennium Zimbabwe enjoyed the unenviable status of having the world's fastest shrinking national economy. Against this unpromising backdrop, joblessness, economic informalisation, and uneven development are unifying and parallel threads across the seven countries that are the geographical focus in this volume.

Over the past two decades, whilst research by economic geographers in these countries has made considerable progress, there has not appeared so far a comprehensive volume which addresses the considerable spatial and economic shifts that have been in evidence recently. Indeed, contemporary review articles concerning the 'state of the art' of local economic geography point to the absence of book-length treatments of the region's changing economic geographical landscape (Rogerson, 2000, 2002a). To a large extent this gap in the outputs of geographical research is a product of the relatively small numbers of dedicated researchers in the region who would be classified as economic geographers. In this book an attempt is made to draw together the writings of the small group of economic geographers who are working in the region of southern Africa with contributions from a number of British-based geographers with longstanding research commitments to the region.

The challenge in this book is to build upon the foundations provided by earlier landmark studies which documented the making of apartheid geographies and of the beginnings of spatial economic change under political transition (Lemon, 1976, 1987, 1995; Simon, 1998). The focus here is an examination of the important and sometimes dramatic changes that are taking place across these seven countries, particularly during the first decade of democratic (post-1994) rule in South Africa. The local changes wrought in the economic geography of southern Africa are a function of both global and regional forces. Undoubtedly, the shifting spatial economic landscape of southern Africa is, in part, a critical function of South Africa's shedding of its pre-1994 status as pariah state and of the country's re-positioning in the global economy (Lemon, 2000). Adjustments to and responses towards globalisation are a recurrent theme across the seven countries that are investigated. But, the unravelling of apartheid South Africa has generated also a set of new regional economic relationships between the neighbouring states and South Africa itself (Gibb, 1998, 2001). Post-apartheid South Africa's acceptance as part of the wider Southern African Development Community has been accompanied by the moulding of new forms of regional economic integration. In addition, for economic geographers one of the most dramatic spatial manifestations of these changes in the regional environment is the strengthening of a range of cross-border development initiatives which are taking root across and between the different countries that are investigated here.

Southern Africa in the Global Economic Environment

The march of globalisation has strongly impacted upon the re-shaping of the economic geographies of South Africa and its neighbours, not least by producing a growing macro-economic convergence throughout the seven countries as several adjustments are made to the realities of a new global economic order. During the 1980s and 1990s many of the governments of South Africa's neighbours were required to implement measures for structural adjustment, which ostensibly were aimed at stabilising their economies as a condition for accessing development finance or credit for servicing old loans from international financial institutions (Lambrechts,

2001). These global forces precipitated a marked convergence in the economic directions pursued by different states, such as Botswana, Lesotho and Swaziland which were charting a path of 'dependent development', and other states such as Mozambique which were engaged in a path of so-termed 'revolutionary transformation'. The case of Zimbabwe under the leadership of President Mugabe is, however, one exception where the country's domestic policies since the late 1990s have flouted the constraints of the global political-economic order. The economic consequences have been devastating both for Zimbabwe's population and economy and more broadly for creating a negative economic sentiment surrounding investment and growth opportunities in southern Africa as a whole. With the exception of Zimbabwe, the majority of southern African governments currently subscribe to macro-economic stability as well as trade and investment liberalisation as the economic mantra for increased domestic and foreign inward investment, and therefore for the enhancement of economic growth.

This situation of macro-economic convergence extends post-1994 even to the case of South Africa, the largest and dominant economy in the southern African region (see Simon, 2001). It must be understood that whilst the economy of post-apartheid South Africa is rapidly globalising as Carmody (2002) shows, 'the nature of that globalisation differs significantly from that experienced by the rest of the region'. Indeed, whilst most countries in southern Africa experienced globalisation as something that was externally imposed and mediated by structural adjustment programmes of the World Bank/International Monetary Fund, in South Africa the momentum for globalisation has been largely internally generated. The post-apartheid state, led by the African National Congress, chose in 1996 to adopt a relatively orthodox macro-economic strategy, the Growth, Employment and Redistribution (GEAR) programme (Republic of South Africa, 1996). For many observers, notably Bond (2001) South Africa's new macro-economic strategy represents a 'homegrown' structural adjustment programme that reflects strongly the influence of the World Bank-led 'Washington Consensus'. Indeed, since 1994 South Africa has introduced a set of highly neo-liberal trade, investment and fiscal strategies which have been interpreted by some researchers as an uneasy compromise between globalisation and social democracy (International Labour Office, 1999; Moore, 2001). With GEAR-led moves towards liberalisation of markets and the ending of sanctions against South Africa, several major South African companies – Anglo American, South African Breweries, Old Mutual and Billiton - have lost little time in further globalising their operations (Simon, 2001). Africa provided the platform or launch pad for the global expansion of many of these enterprises. Since 1994 many of South Africa's leading conglomerates, despite an 'improving business climate', have de-linked from the country, moving their primary stock market listings from Johannesburg to London. Overall, therefore, South Africa represents an unusual case of globalisation which has been instigated 'from the inside out' (Carmody, 2001, 2002).

The impact of globalisation in South Africa merits close attention in light of the country's status as regional superpower and of its economic dominance throughout the region. Despite the faith squarely placed in the tenets of market liberalisation since 1994 and of redistributing resources towards the poor, it is evident that, as one recent commentator remarks, 'the market has not fully returned the favour' (Lamont, 2001, p. i). As Pillay (2000, p. 7) observes 'even the most fervent supporter of GEAR would concede that it has had limited success in attaining all its goals'. 'The most pressing problem facing South Africa today', according to one World Bank investigation, 'is the absence of sustained economic growth and job creation, which are essential to reduce poverty and improve living standards' (Lewis, 2001, p. 1). An ILO (1999) study records that the employment situation has deteriorated and that trade liberalisation may have shifted production in favour of capital-intensive sectors and to the detriment of labour-intensive activities. Despite the elimination of international sanctions, since the beginnings of neo-liberal economic reforms in 1996 more than half a million jobs have been lost in the formal economy of South Africa in contrast to the 600,000 that were targeted to be created by the GEAR reforms (Carmody, 2002). Although the brunt of employment losses have occurred in mining and construction, the decline in employment in manufacturing, the largest contributor to South Africa's GDP, is 'a cause for serious concern' (Bhorat, 2001, p. 7). Overall unemployment is presently estimated at nearly 36 per cent, varying from near zero for highly skilled groups of workers to over 50 per cent for unskilled and semi-skilled workers (Lewis, 2001).

With an annual economic growth rate running at less than 3 per cent, the task of turning South Africa into a leading non-racial emerging market economy is proving immensely difficult. As a World Bank study observes: 'the growth-and-employment challenge facing South Africa is a daunting one. Investment rates are low. FDI (foreign direct investment) inflows disappointing and the unfinished agenda of structural reforms leaves South Africa at a disadvantage within an increasingly competitive global environment' (Lewis, 2001, p. i). One recent investigation disclosed that the unemployed have swollen to represent at least 25 per cent of the economically active population with a further burst of (unwelcome) growth taking place in many low-productivity activities surrounding informal employment (Pillay, 2000). The failure of the economy of post-apartheid South Africa to fulfil optimistic expectations contributed to the precipitate fall of the rand, with a 25 per cent devaluation occurring during 2001 to a rate of over R10 to one US dollar. South Africa's disappointing economic record is in part linked to the emerging market crisis of 1998. Several other negative factors must also be noted, including an economic environment subdued by downturns in the economies of the USA and Western Europe, the consequences of the war on terrorism, the 'Mugabe land grab' and political violence in Zimbabwe (chapter 13). With the globalisation of South African enterprises, inward foreign investment flows have been limited and many domestic South African enterprises have postponed expansion plans (Hough, 2000; Lewis, 2001; Carmody, 2001, 2002).

It has been argued that what is missing in both South Africa and the southern African region as a whole are sets of policy initiatives that focus on structural economic change (Lamont, 2001). This focus on structural change and economic diversification is, however, being addressed in certain new policy initiatives, at least in South Africa and Botswana. In Botswana the government is keen to boost economic diversification in the country's minerals-dominated economy and is now promoting efforts to develop a 'knowledge economy' through greater spending on education and by encouraging private sector investment in manufacturing and services. In South Africa, planning is well-advanced for building a knowledge and technology-based economy with the introduction of targeted industrial policies in order to attract both domestic and foreign investment (Republic of South Africa, 2001). Although the new proposals reassert the importance of a stable macro-economic framework they also highlight the need for a 'developmental state' to intervene and 'counter the powerful tendencies toward inequality, uneven development and marginalisation that characterise the globalisation process' (ibid., 2001, p. 3). Essential themes in the new strategy document focus upon supporting 'knowledge-driven activities' with a specific sectoral focus on such promising sectors as information technology, automotives, clothing, electronics, financial services and especially tourism. Indeed, throughout the seven countries of southern Africa that are investigated in this book tourism emerges as one economic sector that is targeted as a potential new and important economic driver.

Changing Regional Relationships and Cross-Border Development Initiatives

Some observers suggest that the increased attention given to concepts such as 'the African Renaissance' and NEPAD may lead eventually to a rediscovery of the 'regional solution' towards the common development problems that confront South Africa and its neighbours (Olivier, 2001). Progress towards regional integration so far has not lived up to expectations, however, with President Thabo Mbeki lamenting that the Southern African Development Community as a regional body is 'way behind' other similar entities. Nevertheless, there are a growing number of smaller cross-border development initiatives which have been founded between the seven countries of the region (Hecht and Weis, 2001; InfraAfrica, 2001; SATCC, 2001). For economic geographers of southern Africa considerable interest centres on the progress of the implementation of two programmes in particular (Figure I.2), namely Spatial Development Initiatives (SDIs) and Transfrontier Conservation Areas, the latter more popularly styled as 'peace parks'. Both these two initiatives accord the state a prominent role in shaping economic development and suggest that it 'can play a leading role through strategic interventions and by providing, together with the private initiatives, the investment necessary to promote capital accumulation and economic growth' (Schoeman, 2001, p. 63). Undoubtedly, these *regional* initiatives are significant new forces for moulding the economic geography of southern Africa. Both are firmly grounded in neo-liberal economic thinking and are wedded inextricably to globalisation, which determines or at least circumscribes the policy instruments available to the region to address its development problems (De Beer et al., 1998; Rogerson, 2001a).

During 1995-96 the Spatial Development Initiatives (SDI) programme was conceived and launched in South Africa as an important component for restructuring the post-apartheid space economy. A number of SDIs subsequently have been introduced and operationalised within South Africa (Rogerson, 2002b). Overall, the SDIs are considered as 'targeted interventions by central government for helping unlock economic potential and facilitate new investment and job creation in a localised area or region' (Jourdan, 1998, p. 718). In this respect, so-termed 'bottlenecks to investment', such as inadequate infrastructure are to be removed, and strategic opportunities for private sector investment are identified (Crush and Rogerson, 2001; Rogerson, 2002b). The architects of the SDIs argue that a 'paradigm shift' occurred in economic policy from the formerly 'protected and isolated approach to economic development, towards one in which international competitiveness, regional cooperation, and a more diversified ownership base is paramount' (Jourdan et al., 1996, p. 2). This stress on regional co-operation linked to global competitiveness is viewed as consistent with international economic trends towards regionalisation and globalisation (de Beer et al., 1998; Rogerson, 2001a).

In implementation, the SDIs have several components (Rogerson, 2002b). First, the 'crowding-in' and co-ordination of public and private sector

Figure I.2 Cross-Border Economic Development Initiatives in Southern Africa

investment in areas of proven, albeit often under-utilised, potential for economic development. Second, ensuring political support, commitment and 'buy-in' for the SDI process from the highest levels of government in order to facilitate a fast and focused planning approach. Third, an emphasis upon promoting public-private partnerships which are facilitated by national government in terms of clearing obstacles to investment and by contributing towards the building of vital of infrastructure (Platzky, 2000; Altman, 2001). Fourth, the use of well-planned and publicised opportunities (such as hosting investors conferences, road shows and the use of electronic and print media) for the private sector to obtain detailed information about the SDIs and packages of potential investment opportunities. A final core issue is clustering key targeted industries around anchor projects in order to harmonise productive activities and thus to seek to maximise their local linkages or economic multipliers (Altman, 2001; Crush and Rogerson, 2001).

The Maputo SDI or Maputo Development Corridor, a joint venture between Mozambique and South Africa, was the first SDI to be established and began operations in 1996. It is admitted that the policy thinking that occurred around the reconstruction of southern Mozambique and of Maputo harbour provided the specific genesis for the SDI concept (Rogerson, 2001a). Indeed, it was at a meeting held in Maputo in 1995 between the respective Ministers of Transport in Mozambique and South Africa, that the two governments agreed to a process which would lead to the establishment of the Maputo Development Corridor. Despite its critics (e.g. Schoeman, 2001), in many respects the Maputo Development Corridor must be judged as highly successful, not least for demonstrating that the SDI methodology represents a means for attracting or leveraging much-needed investment into Africa (de Beer, 2001). Of special interest is the construction of the Mozal aluminium smelter, located just outside Maputo, signalling the re-emergence of Mozambique as an international investment location.

Although many of the first wave of SDIs were confined to exclusive operations within South Africa, following the model of the Maputo SDI, after 1996 a number of other SDI initiatives were launched which were centred on cross-border economic development. The acceptance of the SDI methodology in countries across the southern Africa region is largely the result of the 'kick-start' provided by the demonstration effects of the Maputo Development Corridor (de Beer, 2001). Now, several cross-border SDIs extend co-operation between South Africa and its neighbours (InfraAfrica, 2001; SATCC, 2001). For example, the Gariep SDI seeks to broaden co-operation between Namibia and South Africa in order to open up resource-rich areas of South Africa's Northern Cape Province and southern Namibia, particularly in terms of mining, agriculture, mariculture and tourism. Likewise, the Lubombo SDI is a three-country cross-border initiative between South Africa, Swaziland and Mozambique and designed to stimulate international competitive ecotourism as well as agricultural developments. Overall progress in these cross-border development initiatives is crucially dependent on 'the political will of national elites and on the state of international relations between would-be or logical participants in these projects' (Schoeman, 2001, p. 66). Not surprisingly, therefore, least progress has occurred with those cross-border projects or proposals that involve Zimbabwe. Despite agreements to establish a so-termed 'Trans-Limpopo Spatial Development Initiative' that would straddle South Africa's Northern Province and Matabeleland in Zimbabwe, efforts to exploit the area's agriculture, mining and tourism resources have been put on hold largely as a result of the climate of political instability and violence in Zimbabwe.

The term 'regional SDIs' is increasingly applied to describe another group of planned development corridors that either have been established or are planned using South African SDI advice or assistance, but which operate completely outside South Africa's borders. Illustratively, the planned Beira Development Corridor aims to establish a cross-border economic development corridor linking Zimbabwe, Zambia, Malawi and Mozambique. As this is an area seen as having inherent and under-utilised development potential, the project is geared to mobilise public and private sector resources in a partnership to utilise the region's rich natural resources and to create a platform for economic growth and development. Another key strategic objective is to re-establish and upgrade the transport infrastructural linkages between Malawi, Zambia and Zimbabwe and the Mozambique port of Beira (InfraAfrica, 2001; SATTC, 2001). The best-known regional SDI is perhaps the Okavango Upper Zambesi International Tourism (OUZIT) SDI which is planned as an integrated tourism development strategy aimed at positioning the southern African region 'as the pre-eminent eco- and adventure tourism destination in the world' (SATTC, 2001, p. 46).

The planning of the OUZIT SDI reflects an overlap between the operationalisation of the SDIs and of another form of cross-border economic development in southern Africa, namely that of Trans-Frontier Conservation Areas or 'peace parks'. As Schoeman (2001, p. 64) points out, in spatial terms the two programmes 'are not mutually exclusive' and a growing overlap in planning is to be observed. Within the planning for SDIs, tourism is increasingly significant and focussed squarely upon objectives for economic development, investment promotion and competitiveness (Rogerson, 2001b). By contrast, the emphasis in the peace parks or TFCA planning is much more centred on the natural environment, conservation and ecology. Although issues of job creation and the development of a world-class tourism destination are recognised, the essential focus is upon conservation, biological diversity and linking the relatively unspoilt African habitat that exists in many of the border areas between South Africa and its neighbours. The creation of 'a continuous, almost contiguous, chain of national parks, game reserves and other forms of wildlife estate' (Koch, 1998, p. 54) is envisaged. The TFCAs are essentially driven by a set of international private initiatives (most importantly the Peace Parks Foundation) working with national governments, the private sector and local communities. In such projects as Kgalagadi Transfrontier Park between South Africa, Botswana and Namibia or the Gaza-Kruger-Gonarezhou TFCA between Mozambique, South Africa and Zimbabwe, the prime concern is on environmental conservation and nurturing partnerships between local communities and the environment rather than economic growth *per se* (Ferreira and van Niekerk, 2001; Schoeman, 2001). Given the importance of job creation and the potentially huge potential for tourism growth in these cross-border peace parks, however, their progress is of considerable significance for reshaping the economic geographies of the borderland regions of South Africa and its neighbours (Koch, 1998; Ferreira and van Niekerk, 2001). Moreover, in a wider international perspective, both the TFCAs and SDIs represent important examples of tourism-led development initiatives that are planned to extend seamlessly across political boundaries (cf. Timothy, 2001).

The Structure of the Book

The book takes as its starting point the end of apartheid in South Africa in 1994, an event which presented the region as a whole with new opportunities and new challenges. Given the dominance of South Africa in the region, the eight chapters in Part 1 address key elements of the South African economy. Part 2 consists of six chapters on those countries that actually border South Africa, which are also, among the fourteen states that now belong to the Southern African Development Community (SADC), those which have the closest economic relations with South Africa. Part 3 widens the lens by examining issues of regional integration, trade, aid and investment in the region as a whole.

Charles Mather provides a critical assessment of South Africa's changing approaches to land issues of redistribution, restitution and tenure reform in the context of wider debates between poverty-alleviation and productivist approaches. State policies in the 1990s were over-centralised and slow to deliver, failed to activate the productive potential of the rural sector and lacked the potential to create a class of independent, small-scale family farms. The recent Land Redistribution for Agricultural Development programme, designed to address these issues, reflects a development perspective that relates closely to World Bank thinking which has itself attracted significant criticism. These approaches to land reform are occurring in the context of changes in the regulatory environment initiated by the apartheid government in the 1980s and now accelerating processes of deregulation and market liberalisation in line with the post-apartheid state's neo-liberal policies. Mather examines the impact of these changes on the structure of agriculture in both former white farming areas and former black 'homelands'.

Fishing has been neglected by South African geographers but has been closely studied by historian Lance van Sittert. In this volume he provides a trenchant critique of neo-liberal policies since 1994. The apartheid state failed to manage marine resources in a sustainable fashion as it pursued policies favouring monopoly and Afrikaner capital. Post-apartheid reforms have continued to benefit monopoly capital, albeit with a corporate face somewhat blackened by employee stock option plans and empowerment deals. For black petty capitalists and unorganised labour in the informal and subsistence sectors the changes have been symbolic, with no radical redistribution of access. The ANC government has been constrained by depleted stocks, prevailing anti-statist economic orthodoxy and its failure (until 2001) to gain political control of the Western Cape province.

Mining is another area that has received far too little attention from geographers, especially given its historically dominant role in the South African economy. Paul Crankshaw brings an academic background in development studies and professional experience as founder and publisher of *African Mining* to bear on the changing political and economic pressures confronting the industry and the responses of the mining companies. Structural changes include the replacement of corporate empires with a smaller number of mineral-focused companies that are seeking international acquisitions and linkages, especially in Africa. Crisis in the gold sector has encouraged radical restructuring, whilst other minerals have grown in importance. Greater local beneficiation of minerals could yield substantial benefits to the economy. State policies have exerted pressure for black corporate empowerment and the promotion of small-scale mining, whilst organised labour has enjoyed some success in remoulding structures and attitudes. Other changes in the regulatory environment relate, actually or potentially, to the ownership of mineral rights and to the environmental impacts of mining operations. Crankshaw also discusses mine responses to HIV/AIDS and its links with mine migrancy.

The reinvigoration of South Africa's stagnant manufacturing sector, with its low-productivity growth and history of entrenched import substitution, is critical to South Africa's long-term economic development. Etienne Nel outlines the evolving history of the country's manufacturing sector before examining recent debates and trends, including the controversial role of the Minerals-Energy Complex. Manufacturing is characterised by declining employment and the continued spatial dominance of traditional core areas. Trade liberalisation is a critical factor in reshaping manufacturing geography, while Nel sees high and sustained levels of foreign direct investment as critical to growth in this sector. Since 1994 the government has encouraged the search for investment, export-orientated industry, small business development, the emergence of defined manufacturing clusters and new spatial strategies. However the obstacles to significantly faster growth of South Africa's manufacturing exports are formidable in present global conditions.

Since the end of apartheid tourism has been seen as a sector with the potential to make a major contribution to South Africa's reconstruction and development objectives, although currently the industry is geared largely to domestic customers. Christian Rogerson discusses the major themes in changing national policy frameworks, including current constraints and the concept of 'responsible tourism' embracing community involvement, empowerment and sustainability. He identifies three core issues for detailed discussion: the role of tourism in poverty alleviation and the interventions necessary to facilitate this role; the promotion of new growth and investment, including new initiatives in marketing, training and mentorship, and the introduction of tourist investment incentives; and the restructuring of the spatial structure of tourism with the help of infrastructural development and local-level initiatives.

South Africa's well developed financial sector is one of the key contributing forces to uneven development. Patrick Bond makes a detailed analysis of how the financial system contributed to the geography of apartheid, to South Africa's development trajectory, to the demise of apartheid and to post-apartheid economic processes, arguing that post-apartheid financial processes have exacerbated problems associated with the unequal and inefficient economic geography inherited from apartheid. Minimal progress has been made in achieving the accountability, transparency and universal access to financial services called for in the Reconstruction and Development Programme. South Africa's reintegration into international markets has been characterised by disinvestment and capital flight, the willingness of the Reserve Bank to accept responsibility for banking and corporate exposure in international markets, and outright corruption. Deregulation of local financial markets has limited South Africa's capacity to withstand subsequent economic crises. Bond argues that it is only through the grassroots-based advocacy of civil society movements that secure and more equitable domestic financial relations can be established through policies of financial delinking.

In his second contribution Rogerson investigates the growth, location, key challenges and policy frameworks confronting South Africa's rapidly emerging IT economy. With the notable exception of Dimension Data foreign multinationals dominate this sector. The industry exhibits a remarkable degree of spatial concentration in Gauteng, with a secondary focus in the Western Cape. Support initiatives relate to enhancement of the human resource base and black empowerment, including education and training initiatives, intervention to establish a networked society and improvements in physical infrastructure. Other initiatives aim to expand competitiveness and foster new IT development. Local economic development planners are playing an important role in some places. Rogerson explores a number of policy challenges including lack of critical infrastructure, the 'digital divide' between the information rich and information poor, the need for liberalisation, a stronger entrepreneurial base and specific policy interventions, and human resource development in the face of skilled labour emigration. Current policy frameworks are assessed in the light of these needs.

Meshack Khosa notes the central part played by transport in the liberation struggle, in the context of an apartheid transport geography characterised by lack of basic mobility, access and social integration. In this volume he examines the consultations and negotiations leading to the National Land Transport Transition Act, analysing the changing nature of transport policy since 1994 and, with the help of two Human Sciences Research Council Surveys, examining public perceptions of public transport, tarred roads and street drainage. In reviewing the government's performance in implementing the goals of the Reconstruction and Development Programme, Khosa concludes that middle- and higher-income earners and urban metropolitan residents rather than the poor and rural dwellers are the major beneficiaries of good quality services.

South Africa's six neighbours all share varying degrees of dependence on South Africa. This arises in part from their smaller size, their landlocked position (Botswana, Lesotho, Swaziland and Zimbabwe), and from geographies of transport and trade which originated in the colonial period. Most importantly, they collectively represent a periphery in the regional and global space economy, whereas South Africa is the core of the regional economy but occupies a semi-peripheral or secondary core position in global terms. Botswana, Lesotho and Swaziland have long participated in a customs union with South Africa; Namibia was an involuntary participant under South African rule but has elected to remain in the union since its belated independence in 1990. Mozambique enjoyed strong links with South Africa during the Portuguese colonial period, based on transport, tourism and migrant labour. Since the end of apartheid some of these links are re-appearing in new forms, most notably the Maputo Development Corridor. Zimbabwe, as the most industrialised country in the region outside South Africa, has long been the latter's most important trade partner in the region, but its deteriorating political and economic situation is already having adverse effects on the economy of its much larger neighbour.

Namibia was the last country in southern Africa to become independent, and therefore emerged into a very different global environment from that which greeted its predecessors. David Simon assesses the achievements of Namibia's pragmatic approach in its first decade of independence in the context of an open but narrowly based economy, an arid environment, a small, unevenly distributed population, profound income inequalities and dependence on South Africa. Rapid industrialisation is impracticable, and agricultural self-sufficiency is improbable. Whilst facing political challenges of consolidating democracy and nationhood, Namibia has achieved notable success in consolidating its sovereignty over strategic resources and infrastructure, resolving disputes with South Africa over the Orange River boundary and Walvis Bay, negotiating a stake in the diamond mines, bringing the fisheries fleet under local control and effectively patrolling the country's Exclusive Economic Zone. Although Namibia has been politically stable since independence, it has largely failed to overcome the inherited legacies of South African rule.

Botswana is broadly comparable with Namibia in terms of aridity, area, population size and distribution, but has enjoyed more than three decades of independence and notable political stability. Agnes Muyoki and Michael Darkoh chart the sound macro-economic and fiscal policies that have enabled Botswana to achieve exceptionally successful mineral-led growth. The economic diversification needed to promote social justice and sustained development has proved more difficult to achieve. Livestock production remains the backbone of the rural economy but continues to contribute poorly to the national economy, suggesting the need to diversify the rural economy. Mining, especially of diamonds, remains the most important sector of the economy. It has helped to open up remote areas and contributed greatly to government revenues, but its linkages with the rest of the economy remain weak and copper-nickel and soda ash mines have to be subsidised. Diversification of the economy away from its current dependence on minerals depends on the development of low-volume, high-paying tourism and manufacturing. Manufacturing growth has been disappointing in the past decade, and efforts to diversify and decentralise the sector have not achieved their objectives. New policies embracing a shift to higher levels of skills and technology, with an emphasis on export markets, will demand high levels of competitiveness if they are to succeed.

Lesotho has lacked both the political stability and the mineral wealth of Botswana, and has depended heavily on migrant labour to the South African mines. Anthony Lemon portrays a rural economy beset by problems of human and animal population pressures, declining yields and soil fertility, recurrent droughts and the need for land tenure reform. As mining employment in South Africa has declined, and the rise of sub-contracting since 1990 has partially reversed earlier wage increases, numbers of female labour migrants to South Africa have increased. Economic diversification depends on the controversial Highlands Water Project and a notable growth of manufacturing, especially clothing and textiles, since 1980, although the sector has significant weaknesses. Other problems include HIV/AIDS, declining examination performance in secondary education, and growing poverty and social exclusion. The combination of political instability and the incapacity of the state to improve the welfare of its people raises fundamental questions about the sovereign state as an embodiment of national identity.

Swaziland is alone in the region in retaining a powerful monarchy and resisting multi-party democracy. Miranda Miles-Mafafo emphasises its vulnerability to global and regional variations in economic activity, especially developments in South Africa. There is heavy dependence on sugar cane, whilst the manufacturing sector is largely reliant on processing agricultural and forest products. Manufacturing shares some features in common with Lesotho, but has grown less rapidly, albeit from a somewhat larger base, in recent years. The contribution of mining has declined, with a shift from iron ore and asbestos to coal and diamonds, and tourism has stagnated since a mid-1970s peak. Foreign capital involvement in all sectors has intensified since independence in 1968, with South African capital playing a dominant role. Miles-Mafafo concludes by assessing government development strategies including millennium projects, infrastructural development and privatisation.

Lovemore Zinyama and Daniel Tevera seek to explain the socio-political and economic factors underlying the economic and political crisis facing Zimbabwe since the late 1990s. They begin with a detailed discussion of population and environment including water resources, wildlife and forests, and the country's rich and diverse mineral base. Discussion of the settler-colonial economy centres on land alienation, industrialisation and racial patterns of urbanisation. In the first decade of independence in the 1980s provision of health, education, infrastructure and agricultural support for the African population was greatly improved in both urban and, especially, rural areas. Remarkable progess was achieved in the smallholder agricultural sector. These policies proved financially unsustainable and many were reversed under the Economic Structural Adjustment Programme (ESAP) from 1990/91. This has also led to moves to privatise or commercialise many of the parastatals established during the 1980s. ESAP has been widely blamed for the marked decline in living standards in the 1990s; the poor, and even many urban middle-income households, have adopted a variety of coping strategies. Zinyama and Tevera conclude with an analysis of the land reform issue in its historical context, explaining the political context of the 'fast-track' programme and the nature of Zimbabwe's crisis which is, at the time of writing, seriously affecting the economic prospects of the wider region.

Graham Harrison's analysis of Mozambique necessarily begins with the socio-economic contradictions of a colonial state whose economy existed substantially by virtue of its dependent relations with South Africa and colonial Zimbabwe. He then examines the key features of Frelimo's socialism in the context of a changing regional political economy and the destabilising activities of Renamo. Mozambique's 'new market geography' followed an early period of liberalisation beginning in 1984 and followed by the *Programa de Reabilitação Económica* (PRE) from 1987 onwards. The PRE was implemented during a civil war, and its implementation both destabilised the Frelimo state and provoked popular resistance. Economic recovery followed in the 1990s, at least in terms of macro-economic aggregates, helped by the end of the war, the World Bank and other donors, and increasing levels of foreign direct investment (FDI). Patterns of growth in different economic sectors reveal contradictions and continuities from previous periods, including a southern concentration reflecting the strength of re-integration with the South African economy and increased socio-economic differentiation. Dependency on external donors and creditors is reflected in a 'hollowing out' of the Mozambiquan state.

Richard Gibb begins his analysis of regional economic integration by reminding us that southern Africa is but a small and peripheral part of the world economy in which the human consequences of marginalisation are profound. Regional integration is seen as a solution, but can only be as strong as the sum of its parts; its proponents are divided between advocates of interventionist and market-orientated, 'open' regionalism. Gibb reviews the achievements of the earlier Southern African Development Co-ordination Conference, its replacement by the Southern African Development Community (SADC) in 1992, and the structures and programmes of SADC including the establishment of a Free Trade Area. The latter is likely to enhance intra-regional inequalities, despite an asymmetrical approach to trade liberalisation. The more territorially limited Southern African Customs Union (SACU) has its origins firmly embedded in colonial policies; Gibb discusses the original 1910 agreement, the revised agreement of 1969 and the outline framework of a new SACU negotiated in 2000 which addresses issues of inequality in an interventionist manner. In the region as a whole, however, the principal obstacles to regional economic integration will continue to arise from inequalities in levels of development and the very high ratio between South African exports to the rest of the region and imports from it.

In the final chapter Lemon and Gibb view South and southern Africa in a global context. They begin by tracing South Africa's path from protectionism and import substitution to post-apartheid trade liberalisation. Official policies are attempting to maintain and strengthen trade relations on many fronts including the EU, North America, China and the Indian Ocean Rim. South Africa is also voicing the wider concerns of developing countries in the World Trade Organisation and seeking to build common approaches on key issues within southern Africa. Scarcity of capital is a critical problem facing the South African economy, and FDI remains a disappointingly small proportion of total investment. Foreign aid is proportionately far less important in South Africa and Botswana, but more significant elsewhere, especially in Mozambique. The EU has become South Africa's largest trading partner as well as aid donor, but offered South Africa only severely qualified membership of the Lomé Convention. Acrimonious negotiations eventually led to the conclusion of a Trade, Development and Co-operation Agreement including the creation of a Free Trade Area (FTA) which is reciprocal but has a development component reflected in its differential coverage and asymmetrical timing. This has already led to significantly increased trade in both directions, but the trading relationship remains characteristic of relations between core and semi-periphery in the world economic system.

References

Altman, M. (2001), 'Evaluation of the Spatial Development Initiatives', unpublished report prepared for the Development Bank of Southern Africa, Midrand.

Bhorat, H. (2001), *Employment Trends in South Africa*, Occasional Paper No. 2., Friedrich Ebert Stiftung South Africa Office, Johannesburg.

Bond, P. (2001), 'Debates in Local Economic Development Policy and Practice: Reversing Uneven Development and Reactivating the State in a 'Post-Washington' Epoch', unpublished paper presented at the Local Economic Development Workshop for Eastern and Southern Africa, Harare, 29 October.

Carmody, P. (2001), 'The Geography of South African Conglomerate Restructuring', unpublished paper, Department of Geography, University of Vermont, Burlington.

Carmody, P. (2002), 'Between Globalization and (Post) Apartheid: the Political Economy of Restructuring in South Africa', *Journal of Southern African Studies*, vol. 28, in press.

Crush, J. and Rogerson, C.M. (2001), 'New Industrial Spaces: Evaluating South Africa's Spatial Development Initiatives (SDI) programme', *South African Geographical Journal*, vol. 83, pp. 83-92.

De Beer, G. (2001), 'The Maputo Development Corridor', unpublished report prepared for the Development Bank of Southern Africa, Midrand.

De Beer, G., Mmatali, R.A., Mahuman, A.C., Nyathi, S.C. and Soares, F.H. (1998), 'Spatial Development Initiatives and the Future Development of the Southern African Borderlands', unpublished paper prepared for the Maputo Corridor Company Policy Research Programme, Nelspruit.

Ferreira, S. and van Niekerk, A. (2001), 'The Gaza-Kruger-Gonarezhou Transfrontier Park: Miracle or Disaster?', unpublished paper, Department of Geography and Environmental Studies, University of Stellenbosch.

Gibb, R.A. (1998), 'Flexible Integration in the "New" Southern Africa', *South African Geographical Journal*, vol. 80, pp. 43-51.

Gibb, R.A. (2001), 'The State of Regional Integration in Southern Africa', in C. Clapham, G. Mills, A. Morner and E. Sidiropoulos (eds), *Regional Integration in Southern Africa*, Southern African Institute of International Affairs, Johannesburg, pp. 71-87.

Hecht, V. and Weis, C. (2001), *Border Crossings in Southern Africa: Regional Development Through Economic Integration and Road Transport Strategies*, Ruhr University, Bochum.

Hough, J. (2000), 'Strategic Trends, Foreign Direct Investment and Competitiveness in Southern Africa', *Africa Insight*, vol. 29 (3/4), pp. 3-7.

InfraAfrica, Ltd (2001), 'Transport Investment Opportunities in an Emerging Market: Changing Transport for Growth and Integration in Southern Africa', unpublished paper presented at the SADC Transport Investment Forum, Windhoek, Namibia, 24-26 April.

International Labour Office (1999), *South Africa: Studies on the Social Dimensions of Globalization*, Task Force on Country Studies on Globalization, International Labour Office, Geneva.

Jourdan, P. (1998), 'Spatial Development Initiatives (SDIs) - The Official View', *Development Southern Africa*, vol. 15, pp. 717-25.

Jourdan, P., Gordhan, K., Arkwright, D. and De Beer, G. (1996), 'Spatial Development Initiatives (Development Corridors): their Potential Contribution to Investment and Employment Creation', unpublished paper, Department of Trade and Industry, Pretoria.

Koch, E. (1998), '"Nature has the Power to Heal Old Wounds": War, Peace and Changing Patterns of Conservation in Southern Africa', in D. Simon (ed), *South Africa in Southern Africa: Reconfiguring the Region*, James Currey, Oxford, pp. 54-71.

Lambrechts, K. (2001), 'The SADC: a Developmental Profile', in Institute for Global Dialogue, *The IGD Guide to the Southern African Development Community*, Institute for Global Dialogue, Johannesburg, pp. 31-54.

Lamont, J., (2001), 'South Africa: Pace of Change is Undermining Old Certainties', *Financial Times Survey*, 26 November, pp. i-iv.

Lemon, A. (1976), *Apartheid: A Geography of Separation*, Saxon House, Farnborough.

Lemon, A. (1987), *Apartheid in Transition*, Gower, Aldershot.

Lemon, A. (ed) (1995), *The Geography of Change in South Africa*, John Wiley, Chichester.

Lemon, A. (2000), 'New Directions in a Changing Global Environment: Foreign Policy in Post-apartheid South Africa', *South African Geographical Journal*, vol. 82, pp. 30-39.

Lewis, J.D. (2001), *Policies to Promote Growth and Employment in South Africa*, Discussion Paper 16, Informal Discussion Papers on Aspects of the Economy of South Africa, The World Bank Southern Africa Department, Washington DC.

Moore, D. (2001), 'Neoliberal Globalisation and the Triple Crisis of "Modernisation" in Africa: Zimbabwe, the Democratic Republic of Congo and South Africa', *Third World Quarterly*, vol. 22, pp. 909-30.

Olivier, G. (2001), 'Regional Integration and African Revival', *Africa Insight*, vol. 31 (3), pp. 39-46.

Pillay, P. (2000), *South Africa in the 21st Century: Key Socio-Economic Challenges*, Occasional Paper No. 1, Friedrich Ebert Stifting South Africa Office, Johannesburg.

Platzky, L. (2000), 'Reconstructing and Developing South Africa: The Role of Spatial Development Initiatives', unpublished paper presented at the International Conference on Sustainable Regional Development, University of Massachusetts, Lowell, 28 October.

Ramsamy, P. (2001), 'Poverty Reduction: a Top Priority in SADC's integration agenda', address by the Executive Secretary of the Southern African Development Community available at www.sadc.int.

Republic of South Africa (1996), *Growth, Employment and Redistribution: A Macro-Economic Strategy*, Department of Finance, Pretoria.

Republic of South Africa (2001) 'Driving Competitiveness: an Integrated Industrial Strategy for Sustainable Employment and Growth', discussion document of the Department of Trade and Industry Pretoria available at www.dti.gov.za.

Rogerson, C.M (1997), 'Globalization or Informalization?: African Urban Economies in the 1990s', in C. Rakodi (ed), *The Urban Challenge in Africa: Growth and Management of its Large Cities*, United Nations University, Tokyo, pp. 337-70.

Rogerson, C.M. (2000), 'The Economic and Social Geography of South Africa: Progress Beyond Apartheid', *Tijdschrift voor Economische en Sociale Geografie*, vol. 91, pp. 335- 46.

Rogerson, C.M. (2001a), 'Spatial Development Initiatives in Southern Africa: the Maputo Development Corridor', *Tijdschrift voor Economische en Sociale Geografie*, vol. 92, pp. 324-46.

Rogerson, C.M. (2001b), 'Tourism and Spatial Development Initiatives: the Case of the Maputo Development Corridor', *South African Geographical Journal*, vol. 83, pp. 124-36.

Rogerson, C.M. (2002a), 'The Economic Geography of South Africa: International Parallels, Local Difference', in B. Maharaj, J. Fairhurst and L. Magi (eds), *Trends and Contemporary Issues in South African Geography*, HSRC Publishers, Pretoria, in press.

Rogerson, C.M. (2002b), 'Spatial Development Initiatives in South Africa: Elements, Evolution and Evaluation', *Geography*, vol. 87, pp. 38-48.

SATCC (Southern Africa Transport and Communications Mission) (2001), *An Overview of the Southern African Regional Development Corridors*, SATCC, Maputo.

Schoeman, M. (2001), 'Transborder Cooperation for Sustainable Development', *Africa Insight*, vol. 31 (3), pp. 62-70.

Simon, D. (1998), *South Africa in Southern Africa: Reconfiguring the Region*, James Currey, Oxford.

Simon, D. (ed) (2001), 'Trading South Africa: Imagining and Positioning the "New" South Africa within the Regional and Global Economy', *International Affairs*, vol. 77, pp. 377-405.

Spicer, D. (2001), 'African Renaissance', *Engineering News*, 16-22 November, pp. 62-6.

Timothy, D. (2001), *Tourism and Political Boundaries*, Routledge, London.

Part 1
South Africa

1 Land Reform and Agriculture

CHARLES MATHER

Introduction

The year 1999 marked a significant change in land and agrarian reform policy in South Africa. In his opening to parliament new president Thabo Mbeki announced the beginning of a strategy for the 'integrated rural development' of South Africa's most poverty-stricken regions. Although no formal document was released in 1999, Mbeki's state of the nation address in the following year provided additional details on the new policy: he declared that the state had 'carried out extensive and intensive work to elaborate an Integrated and Sustainable Rural Development Programme' (Mbeki, 2000). The focus of the new programme is rural areas in both former homelands and former white commercial farming areas; in the latter areas the goal is to de-racialise the platteland and introduce black farmers as equal participants in modern commercial agriculture. Land reform is regarded as one of the cornerstones of Mbeki's integrated rural development programme and recent far-reaching policy changes announced by the Department of Land Affairs are designed to support the programme.

The context for these changes in policy has been increasing concern over the slow pace of land reform, the spectre of what the local media calls 'Zimbabwe-style' land invasions and persistent and probably worsening poverty levels in South Africa's rural areas. In the period since 1994 the ANC has fallen well short of its promise to redistribute 30 per cent of the country's land to black people. Although the speed of delivery improved after 1998 by the end of 2000 less than 1 per cent of land had been redistributed. Restitution was equally slow: of the more than 25,000 restitution cases, only 9 had been finalised by June 1998. A year later the figure was comparatively better at 237 but still unsatisfactory. Budget constraints are not to blame for these disappointing figures as the Department of Land Affairs has consistently under-spent its progressively smaller share of the national budget. A more recent concern has been the escalation of land invasions in Zimbabwe and a smaller number of similar acts in South Africa. Although the DLA has remained firm on its policy against land invasions, in mid-2000, Gilingwe Mayende, Director General of Land Affairs, admitted that the situation in Zimbabwe had 'sharpened focus on the importance of land reform' (cited in *Business Day*, 5 July 2000). The number of land invasions that have occurred in South Africa is, however, not comparable with that in Zimbabwe, and the state has taken action swiftly against such invasions. A second difference in the South African context is that the most well-publicised land invasions have been *urban*, a stark reminder that tenure insecurity is not restricted to the country's rural areas. Finally, several important and ongoing poverty studies have added urgency to a land reform policy that effectively addresses the legacy of apartheid. The studies have found that poverty in South Africa is an overwhelmingly rural phenomenon: the percentage of the country's poor living in rural areas (poverty share) and the poverty rate in these areas are both over 70 per cent.

This chapter begins by outlining, very briefly, the land reform policy that was established in the period between 1994 and 1999. Recent assessments of this programme, in particular the redistribution component of land reform, are considered. The second section explores how this assessment has re-shaped land reform policy in the context of a new integrated development initiative. In the third section this new approach is considered in the light of previous debates on land and agrarian reform. The final section reviews changes in the regulatory environment and assesses their impact on agriculture in both commercial areas and former homelands.

Reassessing Land Reform

There are three key elements to South Africa's land reform programme: tenure reform, restitution and land redistribution. Tenure reform aims to address the insecurity of households living in a range of different situations including former homelands, informal settlements around urban areas and farm workers on white-owned farms. This aspect of land reform is not only the most complex, it also has the potential to affect as many as 13 million black South Africans. Measures to improve security of tenure in former 'white' South Africa are aimed at protecting the rights of labour tenants and farm workers. However, the Land Reform (Labour Tenants) Act of 1996 and the Extension of Security of Tenure Act (1997) have had contradictory impacts (Williams, 1996b). While both were drafted with a view to securing tenure on white-owned farms, the unintended consequence has been the illegal eviction of many households by white landowners attempting to pre-empt claims. The Department of Land Affairs' (DLA) efforts to secure tenure in the former homelands represented its most carefully considered effort in land reform (Lemon, 1998). Aware of the complexity and political sensitivity of addressing tenure in the former homelands given the demands of individuals and of traditional authorities, the DLA spent two years (1996-98) drafting a land rights bill, which would then be subject to comment and form the basis for legislation. In the meantime Minister of Land Affairs Derek Hanekom enacted an interim measure to protect existing rights in former homelands, and refused to sign any transfer of land within these areas. According to individuals involved in the process, the draft land rights bill would protect individual tenure while at the same time allowing for groups to control and allocate land where there was sufficient consensus (Claasens, 2000). Unfortunately, the new Minister of Land Affairs Thoko Didiza shelved the bill shortly after being appointed to cabinet. Although a new proposal is imminent, at the time of writing no concrete plans on tenure reform have been released. Of great concern to those involved in the land rights bill is that the new approach appears to favour the transfer of land to 'tribes' at the expense of individual demands for security of tenure (Cousins, 2000a). The delay in presenting a comprehensive policy on tenure reform in the former homelands is costly in that insecurity in these areas is 'an underlying cause of poverty and rural conflict' (Adams et al., 2000, p. 111).

The second and third components of land reform - restitution and redistribution - are at first glance less complex (but see du Toit, 2000). Restitution targets households who were forcibly removed since 1913 and claims are handled through the Land Claims Court and Commission. A very high percentage of these claims are for land in urban areas and most of these have been settled by cash compensation (Cliffe, 2000). Land redistribution targets households who could not prove that they were forcibly displaced and allows those who can prove that they earn less than R1,500 a month to acquire land through a R16,000 grant. As noted earlier, both of these aspects of land reform have progressed slowly in the face of increasing pressure from claimants.

From 1997 onwards the DLA attempted to monitor the impact and effectiveness of land reform through its monitoring and evaluation programme. The focus was on how the policy had affected the 'quality of life' of beneficiaries. The result was a more streamlined process and one that was also in line with the World Bank's living standards measurement surveys (for details see May et al., 2000). The results of the survey based on the new framework were assessed by several local researchers and by the World Bank's land reform experts. The assessment - which has been presented in various published and draft documents (Deininger, 1999; Deininger and May, 2000; Deininger et al., 1999; May et al., 2000) - has revealed both positive and negative aspects of land reform. A positive finding is that land reform beneficiaries were found to be poorer than the average for rural South Africa; in other words, a relatively wealthier rural elite had not benefited from the programme at the expense of the rural poor. A second positive finding was that land reform grants had generated a 'significant number of economically successful projects' (Deininger et al., 1999). Indeed, in one of the projects beneficiaries were earning upwards of R10,000 each. Yet the overall assessment of the land reform programme, and in particular the land redistribution component, was overwhelmingly negative. These negative findings, and the recommendations that have emerged from them, have played a key role in reformulating not only land redistribution, but also tenure reform and restitution.

A key concern of the recent assessment of land reform has been the failure to 'activate the productive potential of the rural sector'. Judging from the reports cited earlier, 'productive potential' is defined in terms of agricultural production. Deininger et al. (1999) argue that despite the liberalisation of agricultural markets, which has lowered barriers to entry and increased the amount of land available on the market while at the same time depressing land prices, the transfer of land to successful black small-scale farmers has been disappointingly slow. Several reasons for the failure of land reform project to reach a satisfactory 'productivity potential' were identified. First, the absence of a beneficiary contribution to the R16,000 grant appeared to play a determining role in the economic success of land reform projects. The 'top group of projects' and those with higher profits were also those where beneficiaries made regular cash or in-kind contributions to the land reform initiative. The absence of such a co-participation requirement 'has given rise to a whole generation of generally unsuccessful large-scale projects where enterprising individuals managed to enlist hundreds of beneficiaries who were neither interested in agricultural cultivation nor had much knowledge about the land reform process' (Deininger et al., 2000, p.14). Based on these findings they conclude that co-participation would play an important role in improving the economic success rate of individual projects and the overall productivity potential of land reform.

A second criticism of the land redistribution programme, which also emerges from the 1999 survey, is concerned with the single R16,000 grant. The survey results indicate that beneficiary projects may be divided into two broad types: there are 'productive' projects, which are economically successful and sustainable, and there are 'subsistence' projects that assist the poor. A single undifferentiated grant of R16,000 is not appropriate for either group: while 'subsistence' beneficiaries were hampered by a great deal of bureaucratic red tape in accessing the relatively small amount of

R16,000, this same amount was insufficient for individuals or groups attempting to use land reform to become productive farmers. Agricultural economists have pointed out how the nature of the grant system has not supported the needs of small-scale farmers. Thus Kirsten (2000) comments that it is 'doubtful whether the programme, as implemented until recently, has the ability to create the class of independent and small-scale family farms that was initially envisaged'. Based on this observation Deinenger and May (2000) recommend that the grant system be restructured to the needs of different beneficiaries.

The final criticism of land reform - the overly centralised and bureaucratic nature of the application and approval process - has been recognised for some time. Claims for land redistribution have had to start with Provincial Departments of Land Affairs who assist in developing a business plan. If a provincial land affairs committee approves the project, they are sent to the Minister of Land Affairs for a final decision. The bureaucratic requirements involved, and in particular the need for central ministerial approval, have meant that the average approval period is more than fourteen months. An obvious remedy to this problem is the decentralisation of decision-making to the local level to improve both delivery and the 'after-care' of land reform projects.

The recommendations for policy change in land reform flow in a relatively straightforward way from the survey of living standards of land reform beneficiaries. A restructured land redistribution programme should require a beneficiary contribution, it should be sensitive to the different needs of beneficiaries and the bureaucratic processes involved in applying for grants should be streamlined and decentralised. As we shall see in the next section, these recommendations have played an important part in reshaping the role that land reform will play in a broader rural development initiative.

Towards a 'Productionist' Land Reform

The transfer of power from Nelson Mandela to Thabo Mbeki in 1999 marked an important shift in the African National Congress' approach to rural development and land reform. The decision to prioritise rural areas for special attention was partly a response to the various poverty studies undertaken during the 1990s which all painted a depressing picture of the state of underdevelopment in rural South Africa. Increased pressure from civil society, most notably through the Rural Development Initiative (RDI), which was launched in Bloemfontein in early 1999, may have also played a role in the new priority towards rural development. Comprised of a group of rural NGOs the RDI's goal was to push rural development issues more forcefully on to the national agenda (Greenberg, 2000).

To the frustration of NGOs, details on the integrated development programme remained unclear during most of 1999 and 2000 (Greenberg, 2000). Although Mbeki provided some additional detail on the programme in his state of the nation address in 2000, the release of the policy document was delayed until November 2000. Titled the Integrated Sustainable Rural Development Strategy (ISRDS) it is described as an ambitious policy to kick-start development in rural areas based on the government's own experience since the end of apartheid as well as international lessons on tackling rural poverty.

Supported by a reformulated land reform programme, the IRSDS is aimed at bringing together key stakeholders in commercial agriculture with a view to making it both globally competitive and racially diverse. The changes in land reform linked to this new initiative have been presented in the DLA's new Land Redistribution for Agricultural Development (LRAD) programme, described as a sub-programme of land redistribution. It is designed to provide 'black South Africans with access to land specifically for agricultural purposes'. Breaking with the previous grant system, which was pegged at R16,000, the new programme provides grants that cater for subsistence farmers (called safety nets), and three commercial farmer windows for small-, medium- and large-scale commercial farmers. Farmers entering at the lowest window are expected to advance towards becoming large-scale commercial farmers. The grant for subsistence farmers remains at R16,000 while the grants for full-time farmers range from R20,000 to R100,000. In order to qualify for support, beneficiaries are required to provide investments of at least R5,000, which may be made in cash or in kind. Higher beneficiary contributions result in higher grants from the Department of Land Affairs.

LRAD's primary goal is to stimulate growth from agriculture. By encouraging the productive use of agricultural land the programme designers hope to create stronger linkages between agriculture and other rural non-farm activities thereby creating jobs and opportunities for entrepreneurs. The grants are not designed for 'purely residential purposes' and are therefore not aimed at addressing the complexity of land hunger in South Africa that has been revealed by recent urban land invasions in both Johannesburg and Cape Town. Indeed, the DLA has indicated its interest in shifting settlement grants for residential purposes in urban and rural areas to the Department of Housing (DLA, 2000). Nor is it aimed at part-time farmers: beneficiaries must be full-time farmers in all but one category (the food safety net grant). Qualifying for a grant is as complex as before in that the approval of grants is based on 'the viability of the proposed project, which takes into account the total project costs and projected profitability' (DLA, 2000, p. xx). Determining the projected costs and profitability of individual projects is likely to require assistance from a consultant, the costs of which will be carried by the grant applicant.

LRAD will have a significant institutional impact in that it will bring the former departments of land and agriculture closer together. This is a 'jointly shared programme' where both departments will be responsible for policy and training issues associated with it. Demand for agricultural extension and other agricultural services will be far greater; indeed, LRAD will require the former department of agriculture to provide a special set of programmes to support beneficiaries under the new programme.

Despite its status as a sub-programme of land redistribution, this new grant system has important implications for tenure reform and restitution, both of which will be 'enhanced under the LRAD'. This implies that all 'claims involving the restitution of rural and agricultural land are now being approached from a developmental perspective, whereby the beneficiaries do not only receive land but are assisted with the necessary infrastructure, input, skills and know-how' (Mayende, cited in *Business Day*, July 5 2000). A development focus has also affected tenure reform in that demands for security of tenure will have more chance of success if they are made together with an LRAD grant application. In practice this means that restitution claims and tenure reform applicants will be eligible for grants as long as these show an 'intention to farm or enter into agri-business' (DLA, 2001).

LRAD is surprisingly specific in the types of operations it envisages supporting given the diverse environmental conditions that exist in South Africa and the recent economic fortunes of white commercial agriculture. For cropland the minimum farm size that will be supported is 40 hectares; the expected minimum income must be at least R60,000. While the size of the smallest horticultural farm is only 15 hectares, the gross income per annum should be at least R300,000. Cattle and other livestock farms that will be supported must not only conform to the size and income criteria, they must also have a minimum number of livestock units.

The release of LRAD was closely followed by a critical response from the National Land Committee (NLC) which argued that the new policy ignored the rural poor. NLC representatives argued that to access the minimum grant, a matching contribution of R5,000 was required, an amount that would effectively rule out participation by the poorest of the poor. While the DLA's new land reform programme did not earn the support of rural NGOs, it did find support from Agri-SA, the renamed former whites-only South African Agricultural Union, and the National African Farmers Union. In a statement Agri-SA argued that the NLC was ignoring the necessity of transferring agricultural land in such a way that farming would remain a 'vibrant sector' of the economy.

Setting the Agenda, Again?[1]

For several reasons the shift in land reform policy is likely to be welcomed by some involved in policy debates and research on land and agriculture in South Africa. The stronger emphasis on the development impact of land reform in the new policy initiative is likely to receive support. In an assessment of the restitution programme, Du Toit (2000) has argued that while the speed of delivery has been too slow, an equally problematic aspect of the programme has been its design. He writes that restitution claimants are not treated as 'participants' in a process that could improve lives, but rather 'as victims of a loss that must be redeemed' (ibid., p. 75). The result is a programme where restitution and development are viewed as irreconcilable goals (see also James, 2000). Du Toit's (2000, p. 79) recommendation is to ensure that 'restitution is done so as to ensure that it can assist and complement the livelihood strategies of those living on the land'. He illustrates his argument using a restitution case study in which the Port Elizabeth Land and Community Restoration Association (PELCRA) was a key player. Established by a group of urban claimants, PELCRA was used to highlight demands and prevent the sale and development of land under the claim. The significance of PELCRA's claim was that they succeeded in shifting the emphasis of the claim from 'restitution as redemption' to 'restitution for development', a difficult process that involved first working through feelings of loss and anger before moving on to the important task of selecting a tract of land and drawing up development plans. For Du Toit (2000, p. 87) one of the key lessons of the PELCRA case is how the organisation 'questioned the presumed distance between restitution rights and development outcomes which is so often accepted as a given in land claims implementation'. McIntosh and Vaughan (2000) provide a similar critique of the redistribution component of land reform. They argue that the focus of the programme in providing land for the poorest of the poor has not assisted beneficiaries in improving livelihoods or in promoting economic growth.

There has also been recognition that the single grant system may have hampered households from establishing viable agricultural enterprises. Even though land prices have fallen sharply in the recent past the R16,000 grant - even when pooled - has proved insufficient to purchase land and invest in agricultural equipment and inputs. The focus on own participation may also provide beneficiaries with greater ownership of land reform projects than was the case before (McIntosh and Vaughan, 2000). Finally, the decentralisation of decision making to local government and the integration of land reform with other development initiatives will, hopefully, lead to a larger number of economically successful projects.

While the DLA's new redistribution programme may have answered some of its critics, it has at the same time drawn the ire of NGOs and others involved in land issues. In a harsh critique of the direction of land reform Cousins (2000b) has written that it represented the 'gutting' of Land Affairs and the 'de facto takeover of land reform by the Department of Agriculture'. He is particularly concerned about the direction of land reform away from poverty reduction as well as the shelving of the land rights bill by Didiza. There is some justification in Cousins' claim: LRAD gives much greater priority to agricultural production in land reform than was the case in the past and the new approach appears must softer on tenure reform. At the same time the policy links restitution and redistribution much more closely with agricultural development and the establishment of black commercial farmers. The Department of Agriculture appears likely to play a role not only in assisting new beneficiaries in designing land reform projects, but also in providing on-going support through extension on technical and market related issues. The LRAD document notes that the demands on the Department of Agriculture (DA) will be great and it will 'create added urgency for reform of the agricultural extension service' and special programmes for land reform beneficiaries (DLA, 2001, p. 12).

Didiza has defended LRAD by pointing out that it does not abandon the poorest of the poor: they will continue to benefit from the R16,000 grant through equity schemes and the food safety net programme. In her budget speech to parliament in 2001 she argued that the new policy deals with one of the major shortcomings of previous land redistribution efforts, namely their inability to 'address the needs of those who were aspirant farmers' (Didiza, 2001, p. 3). Didiza goes on to say that the new approach effectively bridges the long-debated gap between poverty-oriented and productionist-led land reform programmes:

> The question we need to ask ourselves is whether such a debate should continue as if what is at issue is a matter of choosing the one (production) and rejecting the other (poverty); or should our approach in this dialogue be to find an expression in both programmes that addresses our national land reform agenda? (Didiza, 2001, p. 4).

Easing the tension between production and poverty relief may not be solved in a programme that entrenches this difference by providing policies for 'commercial' and 'subsistence' beneficiaries. One reason why this may be the case is the long and largely unsuccessful history of state polices which have differentiated productive and welfare beneficiaries in many settler societies. In South Africa this particular dualism was manifested in former homeland areas with programmes to improve household food security, while organisations like the Development Bank of Southern Africa attempted to establish black smallholders on 'viable units of land' through its Farmer Support Programme (Watkinson, 1996). Homeland agricultural parastatals also established capital-intensive irrigation schemes, most of which were never profitable.

A second problem with the dualism between productive and welfare beneficiaries has been challenged by a series of innovative studies on the economic value of crop production, livestock production and natural resource harvesting in rural South Africa (Shackelton et al., 2000). The authors estimate that the total value of these land-based activities may be over R5,000 per household per annum, clearly a crucial income source for households that have had difficulty securing employment in the formal sector of the economy. For Shackelton et al. (2000) the implication of this research is that these activities should be recognised and that ways should be found to augment and enhance them in a way that is complementary to commercialisation. They warn against neglecting 'opportunities to enhance current livelihood strategies by a single-minded focus on only those producers who are fully market-oriented or on production systems and products which are already fully commercialised' (ibid., 2000, p. 60). With its emphasis on commercial production for formal markets LRAD appears to favour modern agriculture while relegating subsistence production to 'food safety nets'. The problem with this, as Cousins (2000b) has argued, is that while 'agrarian change' is perceived as a developing a class of emerging farmers, land reform is relegated to a welfare project.

Finally, LRAD's emphasis on the establishment of black commercial farmers means that it is likely to favour wealthier rural households. At least one reason why this may be the case comes from the 1993 World Bank poverty study. The results of the national survey found that a very high percentage of South Africa's rural population depended on agriculture as a source of livelihood. There was, however, an important difference between subsistence farmers on the one hand and farmers who were making money from agriculture. The latter, a much smaller percentage of the rural population, also had a second source of regular income (Schirmer, 2000). That this should be the case is not surprising: it is well known that white commercial farmers who retain access to the land during difficult periods due to markets or the weather, do so through investments in non-agricultural enterprises. Successful white and black commercial farmers in South Africa both appear to diversify their 'businesses' into other sectors of the rural economy in order to spread their risks away from agriculture. The implication of this for LRAD is that better resourced black households are more likely to benefit from a restructured grant system, which demands significant co-participation. While the 1999 survey indicated that land reform beneficiaries were poorer than the rural average (May et al., 2000), future surveys may find a larger number of wealthier rural residents benefiting from the policy.

If the new approach to rural development and land reform benefits better resourced rural households, the question around its impact on poverty alleviation remains unanswered. As noted earlier, Didiza has defended the policy by arguing that the R16,000 will remain in place for the poorest of the poor. The LRAD document also appears to suggest that the growth of the agricultural economy will provide additional new opportunities for the rural poor. The LRAD document indicates the importance of 'simulating growth from agriculture' and 'creating stronger linkages between farm and off-farm income-generating activities', statements which suggest that policy makers expect employment opportunities to be generated outside of agriculture. In other words, poor rural people will benefit indirectly from a more vibrant, efficient and deracialised commercial agriculture (cf. Zimmerman, 2000). There are two likely sources for the idea that growth and poverty alleviation is possible through agriculture. The first is a series of studies on the role of agriculture in the economy. During the late 1980s and early 1990s agricultural economists spent considerable effort exploring the economic role of agriculture in the South African economy. Their results suggested that its indirect role in the economy - through backward and forward linkages - was much greater than might be expected. Based on data from the late 1980s, van Zyl and Vink (1988) found that for every R1m. invested in agriculture, an additional R600,000 was generated in other sectors of the economy. Investment in agriculture also appears to generate more jobs than similar investments in other sectors of the economy, although no mention is made of the quality of these employment categories in terms of wages and conditions. A second source for the idea that agriculture can generate opportunities for the poor comes from the work of Michael Lipton. Drawing on the experience of rural development in East Asia he argues that new opportunities for employment will be 'abundantly generated around small or labour intensive farming' in South Africa. These opportunities will be primarily around construction, transportation, services and trade (Lipton et al., 1996). There is, however, some doubt that the East Asian experience can be replicated in South Africa (Hart, 1996).

A great concern for those who have been involved in debates on land and agrarian reform for longer than a decade will be the striking similarity between LRAD and the World Bank's *Options for Land Reform and Rural*

Restructuring in South Africa produced in 1993. In his detailed assessment of the World Bank's influence in South Africa's land and agrarian reform programme, Williams (1996a) examines how its early thinking focused on the establishment of 'yeoman farmers' on economically viable farming units. In the South African context it envisaged a process that would involve the removal of subsidies for white large-scale agriculture, which would result in the removal of inefficient white farmers. The new land made available through this process would not only decrease land prices, it would also provide opportunities for new smaller-scale black commercial farmers. These farmers would be selected 'based on their prior farming skills and their ability to pay for part of the land cost'. Although the scheme would not 'directly benefit the poorest', they would benefit from employment opportunities generated as a result of the construction of infrastructure. Williams (1996a) argues that by 1993 the influence of South Africans with 'first-hand knowledge' of land and agrarian issues had shifted the World Bank's thinking into accepting a programme that would target poor communities who, through a range of different collective associations, would farm the land productively. The result of this change in thinking was the policy that emerged in the mid-1990s focused on a R16,000 grant for poor rural households.

In assisting in the reformulation of land reform, World Bank advisors appear to be arguing that while liberalisation had the intended effect on white commercial agriculture - more productive land on the market for sale, lower land prices and the removal of previous racially discriminatory 'market distortions' - the instrument to assist commercial agriculture was inadequate. In other words, while all the conditions for the emergence of a class of black commercial farmers were met, the R16,000 grant was not sufficient for new aspiring entrants to commercial agriculture. While this argument suggests a 'missed opportunity', how sustainable new farmers will prove to be in a commercial agricultural sector - the most productive parts of which are being battered by the impact of deregulation and by increasingly competitive global markets- remains to be seen.

An equally important concern is that the new policy does not appear to have addressed earlier critiques of the World Bank's approach, which is strikingly similar to LRAD, or what people on the ground perceive as the most pressing problems facing poor rural South Africans. As Williams (1996a, p.165) wrote some time ago, 'The World Bank's *Options* sought to square a number of circles: redistributing land and maintaining agricultural production; providing for the poor while settling people with resources to take up land and cultivate it commercially; setting up a national programme yet implementing it at a local level. These dilemmas are not resolved'. It is unlikely that the current formulation has resolved these dilemmas. The focus on the establishment of small-scale farmers and the concern to 'de-racialise' white commercial agriculture also seems to ignore more pressing demands for tenure security in both urban and rural areas. As Lund and Wakelin (cited in Williams, 1996a, p. 165) argued, 'Land redistribution cannot be reduced to the need to create a black farming class. Poverty, and the lack of secure rights, is most fundamentally the condition of the majority of black rural dwellers, whose ability to farm commercially is limited by a risk-aversion lifestyle choice, a desire for security and a severe lack of investment resources'.

Policy Reforms: Deregulating Agriculture

Since mid-1990s agricultural policy in South Africa has been driven by a clearly defined goal: to reform an agricultural system that had created an inefficient and costly system which supported white commercial farmers while at the same time discriminating against existing or aspirant black producers. In line with the post-apartheid state's neo-liberal economic policy it has not, however, turned the tables by shifting support from white to black farmers. Instead the state has liberalised agricultural markets in the expectation that a 'free market' will create a more efficient agricultural sector while at the same time creating opportunities for black farmers. Agricultural economists have played a key role in both supporting this argument and assessing some aspects of the impact of policy changes on South African agriculture (van Zyl et al., 2001). This section seeks to trace these changes and explore their impact on the structure of agriculture in both former white and former homeland areas of South Africa. The impact of deregulation must be read in the context of the state initiatives already discussed which aim to secure a place for black commercial farmers in a sector where owners have been white and workers have been black.

Support for farmers in South Africa was for many years based on the racial and spatial divisions associated with apartheid. Within the former homeland areas, tenure insecurity, overcrowding, poor infrastructure and a poor resource base have hampered agricultural production. The cumulative impact of these processes on agriculture has been serious for residents of former homelands. Since the 1970s income earned from agriculture has declined steadily and by the mid-1980s agricultural earnings were only 10 per cent of household income. More recent poverty studies have confirmed that many rural households continue to rely on remittances and pensions rather than agriculture as a source of income generation. There are important exceptions to this general process of decline. During the 1980s investment by homeland departments of agriculture in agricultural projects focused on horticulture, livestock and dry land production increased significantly. Most of these projects were run as estates and managed by financial and technical experts. Although a few projects were run by farmers independently of any management structure, these were the exception. Very few of the projects were economically sustainable or profitable despite the significant economic resources dedicated to their establishment and maintenance.

Following the failure of many estate projects during the 1980s, the Development Bank of Southern Africa funded a Farmer Support Programme (FSP) aimed at establishing individual small-scale commercial farmers. Between 1987 and 1992 as many as 25,000 farmers received support at an average cost of R50,000 per farmer. According to a recent report, very few of these farmers succeeded and 'rather than settling farmers on a dynamic path of accumulation the FSP acted mostly as a welfare transfer'. There are numerous assessments of the programme most of which conclude that lack of capital, poor infrastructure and little market power were responsible for the failure of this ambitious programme (Bookwalter et al., 1998).

Large-scale poverty surveys and detailed case studies of farming in the former homelands suggest that the decline of agriculture in former homelands should be spatially disaggregated. In the former KwaZulu, Transkei and Ciskei homelands the potential for agricultural production is better and this may be why agricultural production remains important. Within these areas, farming appears to play a different role for poor and relatively wealthier households. The large-scale poverty studies conducted during the early 1990s suggest that for a significant proportion of poor rural households, agricultural production remains an important safety net. At the same time, for a very small percentage of rural households, agricultural production is a source of income; households earning an income from agriculture often have a second source of cash income. Sugar-cane farmers in the former KwaZulu and KaNgwane homelands, many of whom have accumulated significant wealth through agriculture, would fall into this category. In livestock production, however, poorer households no longer appear to have access to this valuable resource. There is evidence from various parts of the country that stock holding is restricted to a small and relatively wealthier rural elite (Schirmer, 2000).

State support for white commercial farmers was based on direct and indirect subsidies, tax relief, support for research and development, disaster relief, price controls and measures to establish marketing boards to co-ordinate both domestic and export markets. From the mid-1980s support for farmers changed its emphasis towards assisting white commercial agriculture in weathering high interest rates and an unsustainable debt burden. To this end the state assisted growers in consolidating debts and also encouraged them to become more efficient. Farmers in marginal maize-growing areas, for instance, were assisted in transforming their farms into pastures for cattle. Under increasing financial pressure, the state also decreased its overall support to white commercial farmers. Indirect support in the form of price subsidies, tax rebates and preferential interest rates was also progressively withdrawn through the 1980s.

In the early 1990s two important reports paved the way for the further liberalisation of South African agriculture. The Board of Tariffs and Trade (BTT) report published in 1992 focused on the association between food price inflation and the structure of marketing. It found that food inflation was disproportionately high and it laid the blame for this situation on the marketing boards. It recommended that the statutory powers of control boards be removed and the price controls be lifted. In response, the Minister of Agriculture appointed a committee of inquiry under Professor Eckart Kassier. The Kassier report, which was delivered in 1993, supported many of the reforms suggested by the BTT. It found that the previous marketing act 'fostered inefficiency and in particular inflated agricultural production and marketing costs, it hampered the management of agricultural risks and it reduced agricultural and agro-industrial income generation and employment' (van Zyl and Kirsten, 2001, p. 32). The impact of these reports played an important role in speeding the pace of reform and deregulation.

In 1996 a new Marketing of Agricultural Products Act was passed by parliament. Key goals of the new legislation include: increasing market access for all participants, promoting efficient production, increasing exports and enhancing the economic and environmental sustainability of commercial agriculture. By 2000, with the notable exception of sugar, all domestic marketing boards had been abolished. A 1993 amendment to the Co-operatives Act allowed these former marketing boards to transform themselves into private companies. Co-operatives taking this route have, however, faced legal challenges from growers who claim - justifiably in most cases - that the resources of the companies that are now in private hands were built up through growers' levies and other contributions by farmers.

The Impact of Reform

Agricultural economists have spent considerable effort tracing the changes that have occurred, primarily in white commercial agriculture, following the implementation of the new Marketing Act. Supporters of free markets, they conclude that the impact of deregulation has been overwhelmingly positive. Several key indicators are used to provide this assessment: first, real food prices have declined or remained constant in the period between 1992 and 1999. While maize prices are currently at the same levels as 1993, prices for other commodities such as meat and livestock products have declined during the 1990s. These data appear to indicate that the former system of marketing was responsible for food price inflation. Second, the deregulation of agricultural markets has had an important impact on land prices; although land prices have been falling since the 1970s, the rate of decline has increased markedly in the 1990s. In certain sectors of agriculture, notably the deciduous fruit industry, farms are being sold for a fraction of their value three years ago. Third, there has been - in the language of agricultural economists - a significant 'private sector response' to deregulation, especially in the deciduous fruit and citrus sectors. When the single-channel exporters of these commodities lost control of fruit exports, a surprisingly large number of agents established themselves as exporters. In 1997 alone 1,800 new enterprises were registered in the agricultural sector in South Africa. Multinational fruit exporting companies including Dole and Del Monte were amongst the new agents who established a base in South Africa to export fruit to the northern hemisphere. A fourth impact of deregulation has been that farmers have become more 'market oriented' and have shifted production from field crops to higher value products in horticulture and livestock production. Within field crops there has been a shift from yellow maize, used primarily as feed for livestock, to white maize, which is the staple of most South Africans.

Deregulation has not, however, proceeded without adversely affecting some sectors of agriculture and indebted farmers and farm workers appear to have been the worst affected. In June 2000 farming debt increased to highest level ever at R30bn. The number of farm bankruptcies also doubled in 1999 as farmers apparently 'dropped out of the free market in agriculture' (*Business Day*, 8 February 2001). With declining land prices, white farmers' debt to asset ratio has increased to almost 32 per cent. Insolvencies have not been restricted to more traditional crops such as maize: the Western Cape's deciduous fruit sector has been in crisis for several years, a situation that is reflected by farm prices which have declined dramatically. The quality of statistics on farm workers has always made it difficult to assess changes in the number of people employed in agriculture. None the less, the data that exist suggest that between 1994 and 1997 as many as 700,000 people were made redundant, mainly from white-owned farms. Howell (cited in Bayley, 2000) provides evidence to suggest that the decline in farm worker opportunities has not been as dramatic. An important research question is the more recent impact of deregulation on the structure and pattern of farm worker employment.

There is much less evidence concerning the impact of deregulation on farmers in the former homelands or its likely impact on new policy initiatives in the land reform programme, specifically the establishment of small-scale black farmers. The evidence that does exist suggests that the few remaining estate projects in the former homelands have collapsed. After 1994 responsibility for the projects was transferred from former homeland departments of agriculture to agricultural development corporations. In the last few years these corporations have been restructured and many are no longer receiving funding from provincial departments of agriculture. In the absence of managerial and technical support, as well as short-term loans, even the most viable projects producing citrus and subtropical fruit have been unable to continue. At the same time, efforts to encourage small-scale farmers by both the Development Bank of Southern Africa and more recently through the Broadening Access to Agriculture Trust (BATAT) have not succeeded. Evidence from the land reform programme's redistribution component also suggests that its efforts in establishing productive farmers have failed.

Farmers who continue to operate in a deregulated environment tend to be better resourced and less encumbered by large debts. Some have been able to adapt to new market conditions by shifting production patterns and investing in new crop varieties. Some of these farmers, both white and black, also have other sources of income to weather droughts and difficult economic conditions. Small-scale farmers, especially those in former homeland areas, have not however benefited either from a land reform programme or from the deregulation of agricultural markets. The impact of deregulation raises serious doubts over the likely success of the state's desire to see black farmers making inroads into a sector of the South African economy that has been traditionally dominated by whites.

Conclusion

Recent changes in South Africa's approach to rural and agrarian reform must be seen in the context of a longer debate over production and poverty and how these are juxtaposed. World Bank advisors to South Africa have always emphasised the productive potential of land reform. South Africa's land reform programme between 1994 and 1999, in contrast, tended to prioritise the impact of land reform on poverty alleviation (Adams and Hollwell, 2000). How these 'key co-ordinates' of the agrarian question are constituted, as Bernstein has argued, reflect 'what is at issue' in the debates on rural development and agrarian reform. His conclusion in 1997 was that 'such debate appears to be politically marginal' (Bernstein, 1997, p.2). While the debate on agrarian reform may not be as politically marginal as before, the way it juxtaposes production and poverty suggests that 'what is at issue' in the agrarian question has changed. Yet the shift towards a productionist land reform is not simply the recognition that land reform should have an economic impact in rural areas. It represents the re-emergence of a particular approach to rural development that is informed by neo-liberal understandings of commercial agriculture and its potential for poverty alleviation in the South African countryside.

Both white and black agriculture have experienced changes in the regulatory environment. State support for white commercial agriculture has declined significantly in recent years and from 1996 new legislation led to the dissolution of marketing boards, some of which had been in existence for more than sixty years. The impact of liberalisation has exposed the vulnerability of farmers who for many years have struggled with huge debts. White commercial farmers with the resources to weather this difficult economic environment have shifted cropping patterns in line with market demands both locally and overseas. Within former homelands, agricultural production has been in decline for some time. The incorporation of former homeland departments of agriculture into provinces and the handing over of projects to agricultural development corporations has resulted in the collapse of agricultural projects, many of which were economically unsustainable despite considerable state support. Those farmers who continue to operate in these vestiges of apartheid geography tend to be those who are better resourced and who often have a second source of income. Agriculture does, however, play a dual role in the former homelands and for a larger and poorer group of households, farming continues to play an important role in sustaining rural livelihoods. Recent state land reform policies are likely to benefit black growers who are already farming profitably.

Note

1.   This heading is borrowed from Gavin Williams' (1996a) important paper which traces the influence of the World Bank on South Africa's earlier land reform programme and policy.

References

Adams, M., Cousins, B., and Manona, S. (2000), 'Land Tenure and Economic Development in Rural South Africa: Constraints and Opportunities', in B. Cousins (ed), op. cit., pp. 111-28.

Adams, M. and Hollwell, J. (2000), 'Redistributive Land Reform in Southern Africa', Natural Resource Perspectives, 64, Overseas Development Institute, Department for International Development, London.

Bayley, B. (2000), 'A revolution in the market: the deregulation of South African agriculture', Policy Paper, Oxford Policy Management, Oxford, England.

Bernstein, H. (1997), 'Social Change in the South African Countryside? Land and Production, Poverty and Power', unpublished paper, Programme for Land and Agrarian Studies, University of the Western Cape, Bellville.

Bookwalter, J., Johnston D. and Schirmer S. (1997), 'The Constraints on Black Farmers in South Africa', unpublished paper presented at the Restructuring of Agriculture in South Africa conference in Durban, April.

Business Day, Johannesburg.

Claasens, A. (2000), 'South African Proposals for Tenure Reform: the Draft Land Rights Bill', in C. Toumlin and J. Quan (eds), *Evolving Land Rights, Policy and Tenure in Africa*, IIED, London, pp. 247-66.

Cliffe, L. (2000), 'Land Reform in South Africa', *Review of African Political Economy*, vol. 84, pp. 273-86.

Cousins, B. (2000a), 'Does Land and Agrarian Reform have a Future and, if so, who will Benefit?' in B. Cousins (ed), op. cit., pp. 1-8.

Cousins, B. (2000b), 'Didiza's Recipe for Disaster', *Weekly Mail and Guardian*, Johannesburg.

Cousins, B. (ed) (2000c), *At the Crossroads: Land and Agrarian Reform in South Africa into the 21st Century*, Programme for Land and Agrarian Studies, School of Government, University of the Western Cape, Bellville.

Deininger, K. (1999), 'Making Negotiated Land Reform Work: Initial Experience from Colombia, Brazil and South Africa', World Development, vol. 27, pp. 651-72.

Deininger, K. and May, J. (2000), *Can there be Growth with Equity? An Initial Assessment of Land Reform in South Africa*, World Bank, Washington.

Deininger, K., Naidoo, I., May, J., Roberts, B, van Zyl, J. (1999), *Implementing 'Market Friendly' Land Redistribution in South Africa: Lessons from the First Five Years*, World Bank, Washington.

Department of Land Affairs (DLA) (2000), *Directorate: Redistribution Policy and Systems: Proposed Plan of Action Following Strategic Direction by the Minister on 11 February 2000*, Ministry for Agricultural and Land Affairs, Pretoria.

Department of Land Affairs (DLA) (2001), *Land Redistribution for Agricultural Development: a Sub-programme of the Land Redistribution Programme*, Ministry for Agriculture and Land Affairs, Pretoria.

Didiza, T. (2001), Land Affairs Budget Vote speech 2001/2002 by the Minister of Agriculture and Land Affairs, National Assembly, 15 May 2001, Cape Town.

Du Toit, A. (2000), 'The End of Restitution: Getting Real about Land Claims', in B. Cousins (ed), op. cit., pp. 75-91.

Greenberg, S. (2000), 'Building a People-driven Rural Development Strategy: Lessons from the RDI', in B. Cousins (ed), op. cit., pp. 361-82.

Hart, G. (1996), 'The Agrarian Question and Industrial Dispersal in South Africa: Agro-Industrial Linkages through Asian Lenses', in H. Bernstein (ed), *The Agrarian Question in South Africa*, Indicator, Durban, pp. 245-77.

James, D. (2000), 'Hill of Thorns: Custom, Knowledge and the Reclaiming of a Lost Land in the new South Africa', *Journal of Peasant Studies*, vol. 31, pp. 629-49.

Kirsten, J. (2000), 'Thoko Didiza is doing the Right Thing', *Weekly Mail and Guardian*, 14 April 2001, Johannesburg.

Lemon, A. (1998), 'Separate Space and Shared Space in Post-apartheid South Africa', *Geography Research Forum*, vol. 18, pp. 1-21.

Lipton, M., Ellis, F. and Lipton, M. (1996), 'Introduction', in M. Lipton, F. Ellis and M. Lipton (eds), Land, Labour and Livelihoods in Rural South Africa, Indicator, Durban, pp. iii-xxx.

May, J., Stevens, T., and Stols, A. (2000), *Monitoring the Impact of Land Reform on Quality of Life: a South African Case Study*, CSDS Working Paper 31, University of Natal, Durban.

Mbeki, T. (2000), State of the Nation Address to Parliament, 4 February 2000, Cape Town.

McIntosh, A. and Vaughan, A. (2000), 'Experiences of Agrarian Reform in South Africa: the Limits of Intervention', in B. Cousins (ed), op. cit., pp.224-34.

Schirmer, S. (2000), 'Policy Visions and Historical Realities: Land Reform in the Context of Recent Agricultural Developments', *African Studies*, vol. 59, pp. 143-67.

Shackelton, C., Shackelton, F. and Cousins, B. (2000), 'The Economic Value of Land and Natural Resources to Rural Livelihoods', in B. Cousins (ed), op. cit., pp. 35-67.

Van Zyl, J. and Kirsten, J. (2000), *Deregulation of Agricultural Marketing in South Africa: Lessons Learned*, Free Market Foundation of Southern Africa, Monograph 25, Sandton, South Africa.

Van Zyl, J. and Vink, N. (1988), 'Employment and Growth in South Africa: an Agricultural Perspective', *Development Southern Africa*, vol. 5, pp. 12-29.

Van Zyl, J., Vink, N., Kirsten, J. and Poonyth, D. (2001), 'South African Agriculture in Transition: the 1990s', *Journal of International Development*, vol. 13, pp. 725-39.

Watkinson, E. (1996), *Enforced Agricultural Change in South Africa: the Emergence of a Small Class of Commercial African Farmers*, unpublished M.A. thesis, University of Natal, Durban.

Williams, G. (1996a), 'Setting the Agenda: a Critique of the World Bank's Rural Restructuring Programme for South Africa', *Journal of Southern African Studies*, vol. 22, pp. 139-66.

Williams, G. (1996b), 'Transforming Labour Tenants' in M. Lipton, F. Ellis and M. Lipton (eds), *Land, Labour and Livelihoods in Rural South Africa*, Indicator, Durban, pp. 215-38.

Zimmerman, F.J. (2000), 'Barriers to Participation of the Poor in South Africa's Land Redistribution', *World Development*, vol. 28, pp. 1439-60.

2 Leviathan Bound: Fisheries Reform in South Africa, 1994-2001

LANCE VAN SITTERT

The Apartheid Fisheries

Central state control of the commercially important maritime commons on the west and south coast of South Africa dates from 1940. Over the ensuing half century to 1990 the state used its power to control access to an increasing number of commercial species and fisheries through the issuing of exclusive use rights. Although dressed up in the discourse of rational scientific management and national interest the real importance of the system lies in its fivefold 'instrument effects' (Ferguson, 1990) on the fisheries. These can be summarised as: monopoly, dispossession, politicisation, delegitimation and resource collapse.

Monopolies in the South African fisheries pre-date 1940, but the state actively assisted their formation in the inshore fisheries after this date through the issuing of exclusive use rights and provision of capital via the Fisheries Development Corporation set up in 1944. By 1953 the inshore sector had become 'heavily trustified'. The 'one plant businesses' characteristic of the pre-war era had been replaced by just four 'financial groups' reaping the 'profits of privilege' bestowed by state largesse estimated to average 25 per cent of owned capital (Union of South Africa, 1953, pp. 58-59, 80-81 and 94).

The transformation of use rights into corporate assets required the negation of all other claims to access. These dated back to the 1890s in the inshore fisheries and were most commonly articulated in terms of communities of fishermen, defined by locale and/or type of fishing, and expressed in the balkanisation of the coast into a host of proclaimed fishing territories. This tradition of local use rights was extinguished in 1940 when Pretoria reintegrated the commons as a unified field for corporate development. Protection for the rights of labour and small capital was initially promised through a minimum wage and co-operatives, but later reneged on by the state.

45

State management unavoidably politicised the fisheries. Access rights were subject to frequent, usually annual, review, allowing the minister wide discretionary power and converting such rights into a lucrative form of political patronage. By the early 1960s National Party credentials and connections were vital to securing access. The state also granted rights to consolidate the support of coloured capital and labour in the wake of Sharpeville (1960), Soweto (1976) and the inauguration of the Tricameral Parliament (1984) (Republic of South Africa, 1986, pp. 79-83). Monopoly capital complained bitterly about the 'bogey of insecure quota rights' hanging on 'the slim end of a ministerial pen' (Oceana, 1961), but was adept at using its political connections to secure the relaxation of conservation restrictions when these impeded profitability.

The state's role in creating monopolies, dispossessing traditional users of their rights and converting access to the marine commons into political patronage had the cumulative effect of undermining the legitimacy of the entire management system. This was most obvious in Namibia where South African authority was never recognised and Pretoria adopted an asset stripping approach to marine and other resources (Moorsom, 1984). South of the Orange, a white electorate and civil society provided some check on the more rapacious instincts of the state and monopoly capital, but the all-pervasive partiality of the system robbed it of any claim to popular legitimacy. Instead, the state found itself assailed on all sides by acts of social banditry, with endemic 'poaching' and the concomitant rise of 'black markets' in the inshore fisheries bearing stark testimony to the illegitimacy of post-1940 state-sanctioned use rights and markets in the eyes of the majority.

Underpinning the loss of legitimacy was the state's obvious failure to carry out its primary mandate: to manage the marine resources in a sustainable fashion. This failure was evident in declining catches, immiserised fishing communities, rising retail prices and increasing restrictions on recreational users. In the first half of the twentieth century the marine commons was portrayed as a frontier of inexhaustible abundance. Fifty years later the discourse was one of scarcity, exhaustion and the threat of imminent extinction of key commercial species. The natural capital of the sea was squandered under the delinquent stewardship of the central state after 1940, fattening monopoly and Afrikaner capital on a diet of fish.

This then was the legacy bequeathed to the post-1994 South African government. Far from abolishing the old order, however, the new nationalists have found it as conducive to their own accumulation and legitimation imperatives as the previous incumbents. The inherited control over access to marine resources has provided them with a rare arena of state action within which simultaneously to facilitate the emergence of black capital and redistribute resources down the social scale. That they have tackled the former with greater urgency and determination than the latter should come as no surprise. The Afrikaner poor were similarly frustrated by their petty-bourgeois nationalist leadership's preference for profitable accommodations with 'Jewish and Jingo firms' (Strydom, 1953) over the pursuit of '*volkskapitalisme*' (O'Meara, 1983).

The Fisheries Reform Process 1994-2001

Any discussion of the post-1994 fisheries needs first to recognise their relative unimportance to the national economy. Despite a landed value of R1.7bn. in 1995, marine fisheries contributed less than 0.5 per cent of national GDP and a mere 1.5 per cent of provincial GDP in the Western Cape (RSA, 1997). Marine fisheries were delegated a provincial responsibility at Union in 1910 and remained so until 1940 in the Cape, while there has never been a dedicated fisheries ministry. Thus there is not, nor has there ever been, a compelling economic or political imperative for effective management of the fisheries. Instead the marine resource has been used to foster ethnic capital, subsidise white agriculture and cement apartheid political alliances.

The ANC promised a break with this past with the 'upliftment of impoverished coastal communities through improved access to marine resources' (ANC, 1994, p. 104), but its priorities were elsewhere. This was evidenced by the awarding of the Environmental Affairs and Tourism ministry to the NP's Dawie de Villiers, with ANC MP and ex-bantustan strongman Bantu Holomisa as his deputy. De Villiers duly initiated consultations on a new national fisheries policy in October 1994 and created a Fisheries Policies Development Committee (FPDC) under the chairmanship of the Food and Allied Workers Union general secretary, Mandla Gxanyana, to oversee the process (RSA, 1996). The FPDC established local and regional Fishing Forums in all the coastal provinces in a bid to involve previously marginalised groups in the policy process. Financial and other constraints, however, effectively sidelined these Forums so that the FPDC was controlled by a 'bosses and workers' alliance between monopoly capital and organised labour in the industrial fisheries, determined to maintain the status quo in defence of corporate profits and members' jobs (Martin and Nielsen, 1997).

The extent of the industrial sector's dominance of the FPDC was revealed in the Committee's recommendations on access rights. Although this was the central issue facing the FPDC it was not debated there, but delegated to a technical committee, which treated it as a scientific not a socio-economic problem (RSA, 1996; van der Elst et al., 1997). This 'grave mistake' 'effectively disempowered' the FPDC (Martin and Nielsen, 1997, pp. 163-4). The technical committee report was highly favourable to the industrial sector, recommending transferable, long-term access rights immune to state interference. The latter was prohibited from withdrawing such rights unless abused or unused and otherwise restricted to 'buy back rights on the open market' (RSA, 1996, p. ii). While allowing that 'new entrants may be preferentially admitted and assisted as part of a phase that attempts to correct past inequalities', the technical committee insisted that 'after this phase, normal market forces should prevail' (RSA, 1996, p. iv). This was deemed necessary to protect profits and jobs from the arbitrary actions of the state.

The FPDC submitted its final report in May 1996. It formed the basis of a White Paper and the Marine Living Resources Bill (MLRB) in 1997 (RSA, 1997a and 1997b). The Bill committed the Minister to balance 'the need to permit new entrants, particularly those from historically disadvantaged sectors of society' against 'the need to promote stability within the commercial fishing industry' in granting access rights (RSA, 1997b, clause 23(3)). To this end, the state proposed both to allocate access rights to a Public Company that would lease them to black small capital and sell such rights by tender to monopoly capital in return for an annual fee (RSA, 1997b, clauses 24 and 28). Only when the total allowable catch (TAC) in a fishery exceeded an agreed maximum could new access rights be allocated, but prior to this, any increase would be shared amongst existing holders. All such rights were to be 'saleable, leaseable, inheritable, divisible or otherwise transferable' and granted for up to fifty years (RSA, 1997b, clauses 23(4) and (8)). This was the industrial sector's notion of 'appropriate' or 'constructive' redistribution of primary benefit to itself (RSA, 1997b, p. 68).

The MLRB's proposed market-driven redistribution – 'tantamount to privatisation of marine resources' (RSA DNA 1998, p. 899) – was vetoed by the Portfolio Committee on Environmental Affairs & Tourism (PCEAT), which reasserted the primacy of both history and the state in the fisheries. The resurgence of the redistribution agenda followed the NP's withdrawal from the Government of National Unity in 1996 and the appointment of Pallo Jordan as Minister of Environmental Affairs and Tourism to replace the departing de Villiers. It was given added impetus the following year by the axing of Bantu Holomisa and his replacement as deputy minister by the arch populist and former chair of the PCEAT, Peter Mokaba. In this radically changed political climate, the PCEAT now insisted on 'the need to restructure the fishing industry to address historical imbalances and to achieve equity' and to this end excised all checks on ministerial power to reallocate access rights from the Bill (RSA, 1997c, p. 2). The minister was now empowered to reduce, reallocate and regulate access rights with 'particular regard to the need to permit new entrants, particularly those from historically disadvantaged sectors of society' (RSA, 1997c, clauses 14, 18 & 24). Such rights were furthermore to be leased not sold and for a maximum of fifteen rather than fifty years. Finally, the proposed Public Company was replaced with a Fisheries Transformation Council (FTC). This was mandated to 'facilitate the achievement of fair and equitable access' through leasing the access rights allocated it by the minister 'to persons from historically disadvantaged sectors of society and to small and medium size enterprises' and the 'development and capacity building' of such recipients (RSA, 1997c, clauses 29-36). As the government reminded the industry, 'the absence of change is the greatest single threat to stability' (RSA DNA, 1998, p. 892).

The revised Bill (RSA, 1997d) was hailed in parliament by Jordan as 'very historic' in guaranteeing redistribution with stability and by the chairperson of the PCEAT as 'a major milestone in reforming fisheries legislation in South Africa' (RSA DNA, 1998, pp. 892 and 897). Another PCEAT member enthused that 'the Bill is a victory for the disadvantaged. It sweeps away the apartheid legislation which resulted in the "larney and boy" mentality which caused so much hardship among all our people, black and white' (RSA DNA, 1998, p. 907). Opposition concerns about the wide powers accorded the minister and accusations that the ANC was attempting to buy the Western Cape vote were swept aside as the Bill passed on a 'wave of euphoria' in March 1998. The Marine Living Resources Act (MLRA) became effective on 1 September 1998 (RSA DNA, 1998, pp. 913 and 922; RSA, 1998). The historical inequalities in the fisheries, however, were to prove far more resilient to change than a triumphant ANC imagined and the wave of euphoria was quickly broken by opposition from the industrial sector.

The Minister's attempt to reallocate a portion of the global TAC in accordance with his new powers under the MLRA was immediately challenged in court by the established industry, which effectively prevented the issuing of any quotas in 1999 (Langklip, 1999). The passing of the MLRA also created rising popular expectations that prompted more than 11,000 applications for access rights, swamping the bureaucracy. Access rights were finally issued for 2000, but, to avoid a repeat of the chaos caused by legal challenges and the popular expectations of 'tata my chance', these were rolled over for 2001 to allow the state time to devise a new system of allocating rights (Kleinschmidt, 2001). Meanwhile the FTC was disbanded in February 2000 amidst scandals over councillors' pay and connections to the industry. These problems, coupled with the ANC's failure to take the Western Cape province in the 1999 general election crippled the MLRA and sapped Pretoria's enthusiasm for state reform of the fisheries. The appointment of Mohamed Valli Moosa as Minister of Environmental Affairs and Tourism in 1999 and his assumption of direct control over fisheries from his discredited deputy Rejoice Mabudafhasi the following year marked the end of the populist moment in fisheries reform and a return to market principles. The subsequent proposals on access rights proposed the scoring of applications strictly in accordance with 'objective' industry-devised criteria and use of stiff application and user fees to restrict the number of applicants. The rights thus granted will be long-term and redeemable by the state only via the market (RSA, 2000 and 2001).

Thus, far from constituting a new deal for the 'historically disadvantaged', the post-1994 reforms have confirmed monopoly capital's dominance of the fisheries and look set to entrench it indefinitely on a soon-to-be privatised marine commons. Despite the rhetoric, the gains made by the 'historically disadvantaged', outside the labour aristocracy of the unions and boardrooms of black empowerment groupings, are token. The post-apartheid fisheries would seem to confirm the old adage that the more things change, the more they stay the same. Indeed, it is the central contention of this analysis that the continuities are far more striking than any discontinuities in the New South African fisheries and that the old 'instrument effects' of the apartheid fisheries continue to operate largely as before.

The Finite Resource

The first and most intractable continuity and the one least amenable to social re-engineering is the marine resource itself, which has effectively set the bounds of the possible for the new government within which any new commercial fish stocks should be (re)distributed, and which must be framed by an acceptance that most have been severely fished down by half a century of sustained industrial exploitation. The inshore fisheries in particular are a pale shadow of the near virgin fisheries inherited by the National Party (NP) in 1948, but awkward questions about past abuses are eschewed in favour of a scientific orthodoxy that fingers environmental fluctuations and so renders history as natural history (Butterworth, 1999).

Indeed, the marine environment has done the ANC government few favours since 1994. The growth crisis in the rock lobster fishery that began in 1989 has persisted, aggravated by a series of mass mortalities induced by toxic plankton blooms (Cockroft and Payne, 1999). The anchovy stock that yielded a TAC of 350,000 tons in 1993 crashed, following a series of poor recruitment years, sending the TAC plunging to zero in 1997, from which it has only slowly recovered (Table 2.1). The poor state of the inshore fisheries has thus denied the ANC 'spare' low-entry-cost resources to allocate to black capital and forced it to confiscate from existing holders instead. This has required the government to try and redistribute white/monopoly capital's share of the TAC and mired it in protracted legal battles for very meagre gains.

Table 2.1 South African Inshore Fisheries Total Allowable Catches, 1989-2000 (Metric Tons)

	Anchovy	Pilchard	Horse mackerel	Rock Lobster Western Cape	Rock Lobster Southern Cape	Abalone
1989	350	30	11.5	4.0	0.45	0.64
1990	150	56	31.5	3.9	0.45	0.625
1991	150	50	24	3.8	0.45	0.595
1992	350	45	24	2.4	0.45	0.60
1993	360	48.6	55	2.2	0.45	0.605
1994	260	43.3	58	2.2	0.45	0.615
1995	210	117	58	2.0	0.428	0.615
1996	70	105	58	1.5	0.428	0.615
1997	0	98	58	1.7	0.413	0.55
1998	175	127	34	1.9	0.40	0.53
1999	-	-	-	-	-	-
2000	123	126	19.4	1.7	0.377	0.5

Source: *Fishing Industry Handbook*, 1989-2000. No quotas were issued in 1999.

Only the offshore fisheries hold out any hope of expansion, but this too is a troubled frontier dominated by powerful monopolies or beyond the control of the state. The hake fishery had recovered from earlier overfishing and was the virtually exclusive preserve of just two companies, I&J and Sea Harvest. The TAC has thus been steadily racheted up since 1994 to achieve the impossible of redistributing access without confiscating that of incumbents (Fig. 2.1).

Figure 2.1 South African Hake Total Allowable Catches 1989-2000 in Metric Tons ('000s)

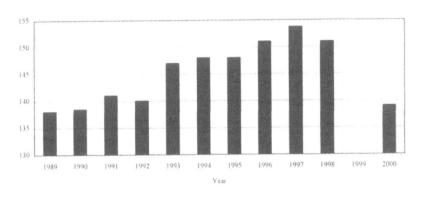

Source: *Fishing Industry Handbook*, 1989-2000. No quotas were issued in 1999.

The market has been less than convinced by this alchemist's trick as evidenced by the declining or stagnant profitability of I&J and Sea Harvest despite the healthy state of the hake resource (Fig. 2.2). Both companies have sought to insulate themselves against the current uncertainty by belatedly taking on black empowerment partners in 1998 and subsequently delisting from the stock exchange (Table 2.2).

Figure 2.2 South African Public Fishing Companies Return on Equity 1989-2000

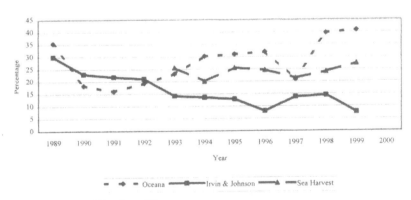

Source: *McGregor's Who Owns Whom.*

Then there are the Patagonian toothfish and a host of other deepwater species touted as targets of future high seas fishing rushes. As the case of the Patagonian toothfish demonstrates, however, prospecting and entry costs are high and the South African state is unable to enforce its control over the fishery in the territorial waters of the Prince Edward Islands, which are consequently plundered by pirate vessels flying flags of convenience.

Caught between the Scylla of inherited inshore dearth and Charibdis of elusive offshore bounty, the new ANC government has found itself in the invidious position of having to preach a doctrine of restraint and the need to negotiate access to the marine commons with the incumbents, chief among which is monopoly capital.

White Skins Black Masks

Although often portrayed as 'a domain for the rich Afrikaners' (RSA DNA, 1998, p. 898), the fisheries have long been the preserve of 'English' monopoly capital, represented by the mining and finance houses that dominate all sectors of the South African economy. Their dominant position in the fisheries has always depended on state largesse making them historically fearful of 'the existence of a populist politician in a senior position' taking 'decisions which are politically satisfying to his personal aspirations ... at the expense of the Industry as a whole' (RSA, 1986, p. 5). Monopoly capital was repeatedly refused greater security of tenure by the populists of the NP and forced to accept a redistribution of access rights to client coloured capital after 1983 and the 'Namibianisation' of the South West African fisheries after 1987 (RSA, 1986; Manning, 1997; Oelofsen, 1999).

It responded to the prospect of an ANC government committed to nationalising the commanding heights with a dual strategy aimed at subverting the policy formulation process (see above), while simultaneously embracing black empowerment (Table 2.2). The first aimed to prevent redistribution by severely curtailing the role of the state in allocating access rights. The second sought black allies (both labour and capital) as insurance against the failure of the first strategy.

This dual strategy initially worked well. The FPDC's proposed market-driven redistribution of access within the fisheries informed the original MLRB tabled in 1997, while a spate of employee stock option plans (ESOPs) and empowerment deals, reinforced by outsourcing and community development initiatives, blackened monopoly capital's corporate face and image. The first strategy suffered a severe reversal, however, when the PCEAT intervened substantially to revise the MLRB and resurrect the interventionist state.

Table 2.2 Black Empowerment in the South African Fisheries, 1994-2000

Year	White Monopoly	Black Empowerment Partner	Share Holding %	Value Rm
1994	Oceana (Old Mutual via Barlows via Tiger Oats)	ESOP	-	-
1995	Oceana	Real Africa	13	66
		Investment Other (Cape interests)	13	-
1996	Irvin & Johnson (Hersov & Menell families via Anglovaal)	ESOP	2.2	23.2
	Oceana	Brimstone	-	7.5
	Sea Harvest (Old Mutual via Barlows via Imperial Cold Storage)	ESOP	8	46
1997	Premier Fishing (Premier)	ESOP	20	-
1998	Irvin & Johnson	Siphumele Investments	10	112
		Ntshonalanga Consortium	5	50
		Dyambu Holdings	5	-
	Premier Fishing	Sekunjalo Investments	60	60
	Sea Harvest	Brimstone	27	148

Source: *McGregor's Who Owns Whom* and *Independent Online.*

The state's subsequent attempt to exercise its power and reallocate access in favour of the 'historically disadvantaged' culminated in monopoly capital successfully interdicting the minister from issuing quotas in 1999 (Langklip, 1999). This historic defeat coupled with that in the national election in the Western Cape (see below), prompted a state climbdown in favour of the market redistribution advocated by monopoly capital. Under the new 'user pays' system, the ability to afford steep application and user fees, not historical dispossession or discrimination, will be key to deciding access. The state will be required to grant access strictly in accordance with published criteria (one of which will be black empowerment) and otherwise confined to verification of applicants' bona fides (RSA, 2000 and 2001). Monopoly capital, it seems, does indeed intend to have its cake and eat it, securing the lion's share of the TAC in perpetuity and decisively defeating its historical *bête noir*, the populist state.

The new neo-liberal *modus vivendi* inaugurated by monopoly capital via the market still excludes the majority of the 'historically disadvantaged', who continue to claim rights of access to the fisheries and look to the state to enforce these rights. Redistribution remains a rallying cry and an as yet unfulfilled promise of the new ANC government.

Bending the Stick in the Other Direction

For advocates of radical redistribution, what is needed is not a 'blackening' of the white monopolies, but their abolition through the state stripping them of their access rights and reallocating these to the 'historically disadvantaged'. As the then Minister of Environmental Affairs and Tourism acknowledged in 1998, there was 'an expectation that the democratic government should now bend the stick in the other direction' (RSA DNA, 1998, p. 896). The radicals have also broadened the scope of the attack to include the state management structure, especially the Quota Board, and the big science practised by its research arm, both of which are seen as defending vested interests. Their call has enjoyed wide support, particularly among black small capital and unorganised labour, for whom empowerment deals and ESOPs have meant nothing except the maintenance of the status quo (Fig. 2.3).

By using the militant rhetoric and tactics of the anti-apartheid struggle, black small capital and its allies, marching under the banner of informal, subsistence or artisanal fishermen, have won a series of concessions. These include involvement in the policy process, amnesty for past transgressions and formal recognition and relief for the subsistence fishing sector (RSA, 1996; Russell, May and Roberts, 1999; Hauck et al., in press).

The insurgency of the self-styled informal sector has not been without its utility for the new government, providing tangible evidence of a popular demand for reform with which to ratchet up the pace of change in an otherwise recalcitrant industry and bureaucracy. The recognition of a subsistence sector has also shifted the centre of gravity in the fisheries away from the politically hostile Western Cape province and drawn the

ANC's traditional heartland of the Eastern Cape into the fisheries patronage network (Russell, May and Roberts, 1999).

Figure 2.3 Redistribution in the South African Fisheries: First Time Entrants (FTEs) 1991-2000 (% Global TAC)

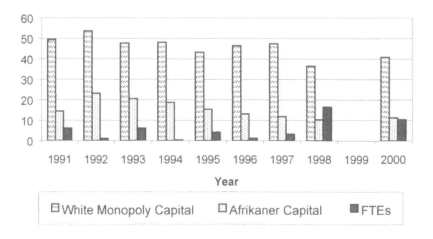

Source: *Fishing Industry Handbook*, 1991-2000. White monopoly capital's share was calculated from Irvin & Johnson, Sea Harvest and Oceana and Afrikaner capital's share from Marine Products and Suiderland. No quotas were issued in 1999.

The informal sector's radicalism, however, is untempered by the pragmatism that is the basic political stock-in-trade of the 'new South Africa' and it has been quick to denounce the new black government and elite in the fisheries as vociferously as the old white nomenclatura. Its radical redistribution agenda is thus ultimately inimical to both the reform agenda of the ANC and the accumulation agenda of black capital. It is also backed up by the threat to subvert the new order through a return to the black economy. That this is no idle threat is borne out by the pillaging of the abalone resource by informal sector poachers in the latter half of the 1990s (Hauck and Sweijd, 1999).

Despite its radicalism and base in the black economy, the informal sector's reach far exceeds its grasp. Whilst some inshore resources are indeed vulnerable to poaching attack domestication has also significantly lessened monopoly capital's dependence on wild stocks in fisheries like abalone (Hauck and Sweijd, 1999). The informal sector simply does not have the capital to threaten the deep-sea resources that are the backbone of the post-apartheid fisheries. It is also a house divided, the term 'artisanal fisherman' obscuring an unstable alliance between petty capital and piece-rate workers under the leadership of the former, who are permanently for sale at the price of legal access. This has once again been paid, with more than half the new entrants to the fisheries since 1994 in the rock lobster (41 per cent) and abalone (11.5 per cent) sectors, where entry costs are low and returns high (*Fishing Industry Handbook*, 1991-2000). Informal sector radicalism has diminished accordingly.

Winning the Western Cape

The ANC government's fisheries reforms have long been interpreted by its opponents as an attempt to secure an ANC majority in the Western Cape province by redistributing access rights. In the run-up to the 1994 election, the ANC certainly attempted to counter the NP's doling out of 'community quotas', by convening a 'fishing forum' under the auspices of ANC provincial leader Allan Boesak's Foundation for Peace and Justice to plan for redistribution in the post-apartheid fisheries (Schutte, 1993; RSA, 1994; Du Plessis and Schutte, 1997). The advantages of incumbency and the cultivation of client coloured capital in the fisheries since the mid-1970s via the Coloured Development Corporation and Drommedaris Fisheries tipped the provincial balance in the National Party's favour (RSA, 1986, pp. 79-83).

Despite the election defeat, control over fisheries patronage shifted decisively to the centre after 1994 with the national government refusing to devolve responsibility for marine fisheries to the coastal provinces and abolishing the Quota Board in 1998 (McQuaid and Payne, 1998). Monopoly capital too showed a decided preference for ANC-aligned capital in its empowerment dealings. The ANC leadership and their business associates in the Western Cape have thus been major beneficiaries of market redistribution in the fisheries. The list includes such luminaries as Franklin Sonn, former South African ambassador to the United States, and the former ANC provincial leader, Chris Nissen, who was installed as chairman of Sea Harvest following the Brimstone empowerment deal.

The fisheries acquired renewed importance in the lead-up to the second general election in 1999 where the ANC again found itself trying to oust an incumbent NP from power in the Western Cape. In this context the PCEAT's revision of the MLRB and the ANC's passing of an MLRA which mandated state redistribution of access were seen by opposition parties as election ploys aimed at swaying the elusive coloured vote. The following year, subsistence quotas were also hastily dispensed ahead of the local government elections. Neither measure had the desired effect, suggesting the limited impact of fisheries patronage, even in the Western Cape. The government's subsequent retreat from the MLRA's redistribution agenda in favour of a free market in access rights is in part a tacit recognition of this (RSA, 2000 and 2001).

A Tenuous Legitimacy

The ANC has dismissed all charges of playing politics with the fisheries by referring to its popular mandate to redress the legacy of apartheid. While true in the broad sense, this begs the question of legitimacy within the much narrower compass of the fisheries, where it is measured not by election results but consent to and compliance with state management measures. Although universal consent and compliance is a chimera, legitimacy, in this narrow sense, has remained as elusive for the post-1994 government as its white predecessor, despite inclusive and exhaustive consultations around fisheries reform.

The lack of legitimacy reflects popular expectations of what an ANC government would herald: for the majority unrestricted access and for the minority the end of the rule of law. An end to all restraint was thus widely anticipated across the political spectrum and during the protracted interregnum prior to the passing of the MLRA in 1998 everyone, from informal sector poachers to recreational users, looted inshore resources at every opportunity. Subsequent strict enforcement has been deeply resented on all sides, both by those for whom the laws are still illegitimate and those for whom the government itself is.

Nor has the formal industry been any more willing to accept ANC hegemony. Monopoly capital fears the redistributive propensities of a populist black government even more than it did that of its white predecessor. Thus, together with its new black allies, it has sought to place its tenure of the marine commons beyond the reach of the state. To this end monopoly capital was willing to subvert the redistribution agenda in the FPDC and, when this was reinstated in the MLRA by the PCEAT, go over into open revolt against the state through the courts. Its victory there sounded the death knell for state redistribution and severely damaged the government's popular legitimacy in the fisheries.

The effect of these relentless guerrilla and conventional assaults on the state's custodianship of the marine commons has been to demoralise both the bureaucracy and government, prompting their grudging retreat in favour of co-management and the market. Although intended to restore legitimacy to state management of the fisheries, the market is itself of doubtful popular legitimacy in South Africa, where it has historically been skewed by the state in favour of the white minority. The ANC, in thus abdicating responsibility for allocating access to the invisible hand, runs the risk of further compromising its already tenuous legitimacy by forcing the 'historically disadvantaged' to look for redress to the not so tender mercies of Mammon.

Conclusion

The ship of post-apartheid reform has floundered on the shoals of history in the fisheries, as elsewhere in the South African economy. The prime beneficiary of the post-1994 reforms in the fisheries has paradoxically been monopoly capital and its junior ESOP and black empowerment partners, for whom the process has delivered long-term access rights immune to redistribution without compensation and a shrunken state presence in the fisheries. For black petty capital and unorganised labour in the informal and subsistence sectors the dividends of democracy have been symbolic rather than real. Recognition and poverty relief rather than the anticipated radical redistribution of access have prompted widespread anger, disillusionment and a return to the black economy (Hauck et al., in press). Finally, the ANC government has found its expected arena of action in the fisheries severely limited by depleted stocks, the prevailing anti-statist economic orthodoxy and its inability to gain control over the rebel ninth province. As a result, the fisheries, which seemed to hold the key to the Western Cape in 1994, have yielded lean pickings in patronage and legitimacy and prompted a ceding of ground to the ubiquitous market.

History, however, suggests that it would be unwise to rush to final judgements on the likely durability of the current neo-liberal order. In 1951, the new NP Minister of Economic Development, challenged by disgruntled Afrikaner capitalists over their continued exclusion from the rock lobster fishery, defended the inherited status quo with a strangely familiar logic: 'I do not like it. It is a virtual monopoly, granted to about 20 of these exporters', but, he quickly added, 'I am not prepared to accept any scheme, even if the present is unsatisfactory, which may have the effect of destroying ... a very valuable national asset' (Union of South Africa, 1951). He therefore dutifully abided by a Fishing Development Advisory Council ruling that excluded small capital from consideration in the granting of new access rights.

Such caution could not and did not last. A decade later, emboldened by two further election victories and with a younger generation in ascendancy within the Party, the Deputy Minister of Economic Affairs, himself a young Turk, bluntly informed the rock lobster exporters association that:

> a redistribution of the rock lobster export quotas was not only fully warranted on fair and equitable grounds but, further it is a wrong approach to consider that a national resource of such value to the exploiter as rock lobster should ... be claimed for all time by the parties placed in temporary possession thereof (Haak, 1962).

He duly confiscated 450 tons from the industry and redistributed it to more than 120 new entrants, mostly Afrikaner small businessmen with political connections to the NP.

It would thus be a mistake to assume that the moment of radical redistribution in the fisheries has passed in the post-apartheid present. It still lies ahead. The economic radicalism of nationalism is after all petty bourgeois, not socialist and its current enforced accommodations with monopoly capital cannot endure the next generation who will contest anew, via the state, the constraint of their accumulation horizon in defence of monopoly profits masquerading as national interest. They may be as cheaply bought as their historical predecessors - both white and black - but this is unlikely, not least because the marine resources are so depleted. In this context the populist tocsin will sound again and the war on monopoly capital - whatever its colour - will commence anew.

References

ANC (1994), *The Reconstruction and Development Programme: A Policy Framework*, Umanyano Publications, Johannesburg.

Butterworth, D.S. (1999), *Taking Stock: Science and Fisheries Management entering the New Millennium*, UCT Department of Communication, Cape Town.

Cockroft, A.C. and Payne, A.I.L. (1999), 'A Cautious Fisheries Management Policy in South Africa: the Fisheries for Rock Lobster', *Marine Policy*, vol. 23, pp. 587-600.

Du Plessis, P.G. and Schutte, D.W. (1997), 'The Socio-economic Effects and Implications of Fishermen's Community Trusts in South Africa', in Hancock, D.A. et al, op. cit., pp. 220-27.

Ferguson, J. (1990), *The Anti-politics Machine*, Cambridge University Press, Cambridge.

Fishing Industry Handbook, 1991-2001.

Haak, J.F.W. (1962), Letter to the Secretary of SAFROC, 11 August, unpublished document in State Archives, Pretoria, HEN1556, 180/29/13 (13).

Hancock, D.A., Smith, D.C., Grant, A. and Beumer, J.P. (eds) (1997), *Developing and Sustaining World Fisheries Resources: The State of Science and Management*, CSIRO Publishing, Collingwood.

Hauck, M. and Sowman, M. (in press), 'Coastal and Fisheries Co-management in South Africa: an Overview and Analysis', *Marine Policy*.

Hauck, M., Sowman, M., Russell, E., Clark, B., Harris, J., Venter, A., Beaumont, J. and Maseko, Z. (in press), 'Perceptions of Subsistence and Informal Fishers in South

Africa Regarding the Management of Living Marine Resources', *South African Journal of Marine Science.*

Hauck, M. and Sweijd, N.A. (1999), 'A Case Study of Abalone Poaching in South Africa and its Impact on Fisheries Management', *ICES Journal of Marine Science*, vol. 56, pp. 1024-32.

Kleinschmidt, H. (2001), 'View from the Helm', *Fishing Industry News Southern Africa*, vol. 2, p. 23.

Langklip See Produkte (Pty) Ltd and Others v Minister of Environmental Affairs and Tourism and Others (1999), (4) SA 734 (C).

McQuaid, C.D. and Payne, A.I.L. (1998), 'Regionalism in Marine Biology: the Consequences of Ecology, Economics and Politics in South Africa', *South African Journal of Science*, vol. 94, pp. 433-36.

Manning, P. (1997), 'Managing Namibia's Fisheries Sector: Optimal Resource Use and National Development Objectives', in Hancock, D.A. et al, op. cit., pp. 696-701.

Martin, R. and Nielsen, J.R. (1997), 'Creation of a New Fisheries Policy in South Africa: the Development Process and Achievements', in Normann, A.K. et al, op. cit., pp. 153-71.

Moorsom, R. (1984), *A Future for Namibia 5: Fishing Exploiting the Sea*, CIIR, London.

Normann, A.K., Nielsen, J.R. and Sverdrup-Jensen, S. (eds) (1997), *Fisheries Co-Management in Africa*, Institute for Fisheries Management and Coastal Community Development, Hirtshals, Denmark.

Oceana (1961), Memorandum: The Rock Lobster Industry and Export Quotas, 27 October, unpublished document in State Archives, Pretoria, HEN1556, 180/29/13(12).

Oelofsen, B.W. (1999), 'Fisheries Management: the Namibian Approach', *ICES Journal of Marine Science*, vol. 56, pp. 999-1004.

O'Meara, D. (1983), *Volkskapitalisme: Class, Capital and Ideology in the Development of Afrikaner Nationalism 1934-1948*, Ravan Press, Johannesburg.

Republic of South Africa (1986), *Report of the Commission of Inquiry into the Allocation of Quotas for the Exploitation of Living Marine Resources*, Government Printer, Pretoria.

Republic of South Africa (1993-2000), *Debates of the National Assembly* (DNA), Government Printer, Pretoria.

Republic of South Africa (1994), *Report of the Commission of Inquiry in Fishermen's Community Trusts*, Ministry of Environmental Affairs and Tourism, Cape Town.

Republic of South Africa (1996), *Fisheries Policy Development Committee (FPDC): Towards a National Marine Fisheries Policy for South Africa*, Department of Environmental Affairs and Tourism, Cape Town.

Republic of South Africa (1997a), *White Paper: A Marine Fisheries Policy for South Africa*, CTP Book Printers, Cape Town.

Republic of South Africa (1997b), *Marine Living Resources Bill (as introduced)*, B94-97.

Republic of South Africa (1997c), *Portfolio Committee Amendments to Marine Living Resources Bill*, B94A-97.

Republic of South Africa (1997d), *Marine Living Resources Bill (as amended by the Portfolio Committee on Environmental Affairs & Tourism (National Assembly))*, B94B-97.

Republic of South Africa (1998), Marine Living Resources Act No.18 of 1998, *Government Gazette*, 18930, 27 May 1998.

Republic of South Africa (2000), *Draft Discussion Document for the Fisheries Management Plan to Improve the Process of Allocating Rights*, Fisheries Management Directorate, Marine and Coastal Management Rights Allocation Unit, Cape Town.

Republic of South Africa (2001), *Stability, Transformation and Growth 2001-2004: The Second Draft Discussion Document for the Fisheries Management Plan to Improve the Process of Allocating Rights*, Fisheries Management Directorate, Marine and Coastal Management Rights Allocation Unit, Cape Town.

Russell, E., May, J. and Roberts, B. (1999), *Subsistence Fishers Task Group Report 2: A Socio-economic and Resource Management Profile of Subsistence Fishers in South Africa*, unpublished report submitted to the Chief Director, Marine and Coastal Management (Department of Environmental Affairs and Tourism), Cape Town.

Schutte, D. (1993), *'n Ontleding van die Ontwikkelingspotensiaal van Geslekteerde Vissersgemeenskappe aan die Wes- en Suidkus*, HSRC, Cape Town

Strydom, G.H.F. (1953), Letter to the Minister of Economic Development, 28 September, unpublished document in State Archives, Pretoria, HEN1559, 180/29/13/37.

Subsistence Fisheries Task Group (2000), *Draft Recommendations for Subsistence Fisheries Management in South Africa*, unpublished report submitted to the Chief Director, Marine and Coastal Management (Department of Environmental Affairs and Tourism), Cape Town.

Union of South Africa (1951), *House of Assembly Debates*.

Union of South Africa, Board of Trade and Industries (1953), *Report No. 337: The Marine Oils Industry of the Union*, Government Printer, Pretoria.

Van der Elst, R., Branch, G., Butterworth, D., Wickens, P. and Cochrane, K. (1997), 'How can Fisheries Resources be Allocated ... Who Owns the Fish?', in Hancock, D.A. et al, op. cit., pp. 307-14.

3   Mining and Minerals

PAUL CRANKSHAW

Introduction

Despite its vast size and its generally conservative response to change, South Africa's mining industry has not been spared the impact of the country's democratisation and re-entry into the global economy. Indeed, few would have foreseen the depth and pace of the changes that have occurred in the past decade as the mines and their owners have grappled with new economic and political pressures. This chapter discusses recent changes in the relationship of mine owners to the various constituencies - domestic and global - that affect how they go about the business of finding and extracting value from minerals.

What was an inward-looking, diversified and staid industry has joined a global investment community that demands higher returns, greater corporate efficiencies and more transparency. The reign of the powerful mining houses (Anglo American Corporation, Rand Mines, JCI, Gold Fields of South Africa, General Mining and Anglovaal) has ended. In the place of these corporate empires are a number of large mineral-focused companies (fewer in number and leaner in capacity), mostly with new leadership and their eyes on international acquisitions and linkages.

New government policies exert pressure for the greater involvement of previously disadvantaged communities, through black corporate empowerment as well as through the promotion of small-scale mining. Organised labour has made its mark in the remoulding of organisational structures and attitudes at mine level, hastening the demise of the previously military style of management. High-level decision-making structures on matters of industry policy have become tripartite bodies in which mine management engages with both government departments and labour representatives.

Mining legislation updated in the last decade has placed greater onus on mine owners for the environmental impact of mining operations, a responsibility enforced further by a growing public awareness of South Africa's natural heritage and potential threats to it.

Shifting the Balance: Labour Demands a Say

It is not by chance that one of South Africa's first, and largest, black independent trade unions took root in the mining sector. The sheer number of workers employed by this sector, the low average level of workers' wages, the poor conditions under which so many of them lived and worked, and the autocratic management style of the mines were all among the grievances seeking redress.[1] The British military model of mine organisation was pervasive in the South African context (Malherbe, 2001, p. 30):

> At the top were the English commissioned officers – the mining engineers, geologists and mine managers. In the middle were the non-commissioned officers – the artisans, miners and shift bosses. At the bottom were the ranks – the drillers and labourers. While there might be some advancement within these layers, the layers themselves were fairly rigid. Certainly the colour bar prevented anyone from the ranks ever advancing above that position.

Figure 3.1 Employment in the South African Gold Mines, 1980-1999

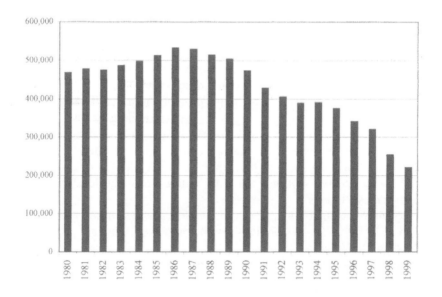

The National Union of Mineworkers' first legal strike was in 1984, setting the tone for the next few years and leading to a much larger industry strike in 1986-87. Mass dismissals were one outcome of this action;

employment in the gold industry peaked in 1986 and has been declining ever since. The Chamber of Mines reported that 534,255 people were employed on Chamber member gold mines in 1986; this was the turning point, and in 1987 the figure slipped to 530,622 (Figure 3.1). By 1990, employment was 474,851; in 1995 it was 377,144 and the average for 1999 was 221,848 (Chamber of Mines, 2000a).

But the more important result was the growing influence that labour began developing in decisions affecting both mine-level operations and issues of a broader nature. The emergence of high-level tripartite bodies to represent the interests of workers, management and government, became a more prominent feature of mining industry decision-making in the 1990s (these bodies dealt with issues ranging from health and safety to training and retrenchment). At the level of the mine itself, committees in the areas of productivity, housing and health and safety are also an integral part of modern management structures.

Gold Gives Way to New Contenders

The dramatic decline since the mid-1980s in both employment and output from South Africa's gold mining industry - once the world's largest by far - have been clear signs of a sector in crisis. Gold output dropped from 605 tons in 1990 to 450 tons in 1999 (Chamber of Mines, 2000a, p. 12), and employment over the same period shrank from 474,000 to 217,000 (Malherbe, 2000, p. 1). Figure 3.2 shows how SA gold production has dropped both in real terms and in relation to global production.

While the dollar price of gold generally declined in the 1980s, the SA Rand devalued faster until 1987, bolstering the price received back home by South African gold producers. This de-valuation was itself the result of crises of a broader nature. It occurred, writes Nattrass, 'most noticeably in 1984 and 1985 when the economy was rocked by the township revolt and subsequent State of Emergency and debt crisis' (1995, p. 857). Then the Rand-price cushion ended and real changes had to be effected to keep mining operations alive. These included retrenchments, high-grading (ceasing the mining of lower-grade areas) and mine closures.

In the face of growing pessimism about gold's future, the transformation of the mining companies and their operations in response to the profit squeeze has been extraordinary, re-establishing the gold sector on a lower cost base with more corporate focus. One industry expert, writing in 1990, forecast that SA gold mining could have as little as eleven years' life left to it, and 22 years at the most. Another suggested that a gold price of $443/oz would be required before the country's gold industry would be profitable (quoted in Nattrass, 1995, pp. 865, 867.) These gold producers are now in realistic pursuit of working costs of $200/oz and under. Working costs among gold producing members of the Chamber of Mines of South Africa were $253/oz in 1999, down from $261 in 1998 (Chamber of Mines, 2000, p. 2).

Figure 3.2 South African Gold Production Compared with the Rest of the World

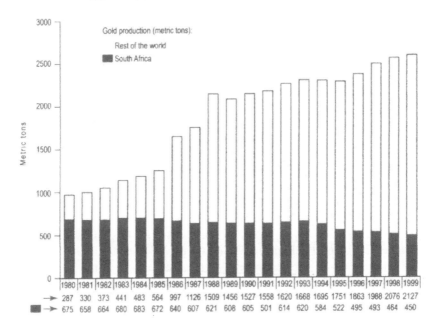

This has been achieved by a combination of labour process restructuring and rationalisation of corporate head office superstructure through down-scaling and mergers. The industry has moved, according to Malherbe, 'to a new model of work characterised by greater worker participation leading to increased responsibility and productivity' (Malherbe, 2001, p. 28). Perhaps the most vigorously endorsed, he says, is the 'managerial' model, in which the mine withdraws from its non-core activities (such as hostel management) to focus on its core business - encouraging flexibility, distributing rather than centralising responsibility, and providing appropriate incentives.

Changes in the nature and size of the corporate vehicles driving the mining industry are discussed in the next section.

While the fortunes of the gold sector have indeed been lean for over a decade, there are other minerals mined in South Africa that are growing in economic importance, and which reflect a more optimistic picture. After all, the country holds the world's largest reserves of manganese (some 80 per cent of global reserves), chromium (68 per cent), platinum-group metals (platinum, palladium and rhodium - 56 per cent), vanadium (45 per cent) and alumino-silicates (37 per cent). It is also prominent in terms of reserves of fluorspar, phosphate rock, titanium, vermiculite and zirconium (Department of Minerals, 2001, p. 4).

Among those minerals recently growing in prominence are coal, of which South Africa produced 223.9m. tons in 1999 (62.8m. tons of this for export) and platinum group metals, of which South Africa produced 46 per cent of world output in 1999 (Chamber of Mines, 2000a, pp. 2-3). The country is also the world's leading producer of vanadium, chrome ore, alumino-silicates, vermiculite and ferro-chrome, and is an important producer of zirconium minerals, manganese, ferro-manganese, diamonds, fluorspar, antimony, silicon metal and iron ore (Department of Mineral & Energy, 2001, p. 4).

By late 2000, investments of some R22bn. had been committed to mineral-related projects in South Africa, of which 80 per cent was for primary minerals and 20 per cent for processed mineral products. Gold and platinum projects made up 54 per cent and 31 per cent respectively of the total for primary mineral projects (Department of Mineral & Energy, 2001, p. 18).

Record high prices for platinum over 2000 and 2001 have led to a renewed interest in the Bushveld Igneous Complex, host to most of South Africa's large-scale platinum mines. Existing mines (belonging mainly to Anglo Platinum and Impala Platinum) are being expanded, and new mines are being started by these and smaller companies (Australia-listed Aquarius and Canada-listed Southern Era Resources among them).

The zinc industry is poised to grow substantially, mainly through the mammoth Gamsberg project in the Northern Cape. Held back in 2001 by a faltering zinc price, this mine, concentrator and smelter complex will involve some R4bn. in capital expenditure. A R2bn. zinc smelter at the Eastern Cape port of Coega is also under consideration.

Easily the largest open-pit mine in South Africa is Iscor's Sishen iron ore mine in the Northern Cape. This sector is being boosted by an R850m. upgrade to the Sishen mine and rail line to Saldanha, with significant beneficiation taking place at Iscor's Saldanha Steel factory.

With two-thirds of the world's chrome reserves, South Africa is steadily increasing its share of the world market in the beneficiated product ferrochrome, meeting 40 per cent of world demand in 1998. Investment in new mines and smelters continues, ensuring that South Africa captures a large proportion of the growth in world demand, currently expanding at a rate of 5-7 per cent per year (Malherbe, 2001, p. 2). Indeed, between 1990 and 1996, employment in the non-gold sectors of South Africa's mining industry actually grew by 10 per cent, giving jobs to 217,000 people.

Global Investment: New Discipline and Direction

The corporate metamorphosis undergone by Johannesburg-based mining companies is as interesting for its causes as its effects. The years of political isolation since the early 1960s, especially the sanctions on finance and investment, had the effect of stemming access to global capital markets and at the same time trapping South Africa's surplus capital inside its borders (Malherbe, 2001, p. 70):

> Perversely, the policy environment encouraged mining houses to invest <u>away</u> from mining. ... Exchange controls prohibited the investment of surplus South African capital in mining ventures abroad; rules 'ring-fencing' mining activities for tax purposes discouraged mining investment in south Africa; and the trade barriers artificially raised returns on import-replacing manufacturing.

Political and economic isolation also cut mining companies off from the changes in how their international peers were managed and how they interacted with their shareholders.

During the 1980s, foreign mining majors, continually under pressure from strong institutional investors, were adjusting painfully to circumstances. Global mining firm RTZ's response to investor demands in the 1980s – focus on mining, lean corporate structure, clear strategy – prefigured the reform of South African mining houses by almost a decade (Malherbe, 2001, p. 70).

Among the South African mining giants' corporate structures, cross-holdings and interlocking directorships were commonplace. Writing in the journal *African Mining* in early 2000, metals analyst Andrew Jones reflected on this aspect of the industry's transformation:

> The arcane and opaque nature of the pre-1997 Johannesburg Stock Exchange Gold Board, with its myriad of listed entities, has gone. The old structure seemed designed only to keep stock analysts in a job and shareholders in a state of confusion, especially those located outside South Africa. The more streamlined structure, consisting of the 'two majors [AngloGold and Gold Fields Ltd] plus a number of essentially local players, means that the South

African gold industry has had to develop well-defined business strategies and communicate these effectively to the investment community as well as its employees. The greater degree of transparency and definition of long-term goals can only be good news all round (*African Mining*, January-February 2000).

Insulation from shareholder scrutiny led to the gradual corrosion of efficiencies in management and the employment of capital. Malherbe (2001) cites an industry leader's opinion on capital-rich mining houses: 'There was little sense of key measures of capital efficiency such as return on assets and return on equity. The companies were technically strong but financially and commercially in the stone age'.

Figure 3.3 shows steadily increasing working costs between 1980 and 1999 and narrow profit margins in the early 1990s before the impact of the South African rand's falling value against the US dollar. Global investors, however, wanted to see exactly where value was being generated, and so wanted better-focused vehicles for mining investment. Of the R500bn. market capitalisation of mining companies on the Johannesburg Securities Exchange, fully 40 per cent of the shareholding is now held by investors overseas. This is a reversal of the trend that saw shareholding building up inside the country's borders in the early years of isolation; Hobart Houghton wrote in the 1967 (second) edition of *The South African Economy* (p. 111) that:

> Ownership of shares in mining companies is increasingly coming to be in the hands of South African nationals and ... the proportion of dividends paid outside of the country is declining. In 1935 only 40.5% of total dividends ... were paid within the country. By 1964 70.9% of total dividends ... were paid within South Africa.

So a painful process of corporate restructuring was inevitable. While the mines themselves bore the brunt of retrenchments in the late 1980s and early 1990s, it was the turn of the city-based head offices in the second half of the 1990s. These structures reflected the traditional strength of the mining finance house, a model that allowed individual mines to draw on a central pool of expertise that they could not sustain on their own. The mining house also supplied much of the capital for new mines and expansions, and the head offices provided the decision-making for these projects.

While Anglo American, like most mining houses, had a division (Amgold) that dealt with its gold interests, the establishment of AngloGold was more significant. With its own board of directors, it was to be operationally independent, and even moved offices to demonstrate an appropriate distance from its previous 'mother company'. The company pursues global aspirations, and is listed in London and New York.

Figure 3.3 Working Costs and Profits in South African Gold Mines

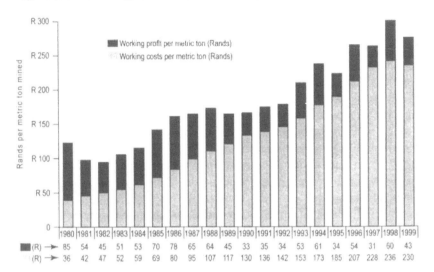

In 1998, Gold Fields of South Africa (GFSA) merged with Gengold (the gold division of Gencor) to form Gold Fields Ltd. The result was the effective disappearance of GFSA and its head office, while the new corporate structure was slimmed down and moved out of Johannesburg city to smaller offices in a peri-urban district. Gencor placed its base metals interests into London-listed Billiton, and operates these under that name out of London offices.

Johannesburg Consolidated Investments (JCI) had its platinum division moved into a new company (Amplats, later Anglo Platinum) by major shareholders Anglo American, and the rest of the company readied for unbundling to black empowerment interests. As discussed later, this vision was never realised and the company's assets were gradually split up and sold off. Anglovaal moved its mining interests into Anglovaal Mining (Avmin), under which is housed its gold interests (in Avgold). The company is still controlled by the Menell family and its future structure is uncertain.

Rand Mines had formed Randgold and Randcoal to house its gold and coal interests respectively, but this was not enough. Randgold was among the first to face a 'palace revolution'; the board was replaced and the company unbundled. Exploration properties in other African countries were spun off into a new London-listed subsidiary, Randgold Resources, which is listed in London. South African gold interests remain in Randgold and Exploration.

In almost all cases, substantial numbers of professional and administrative staff were shed from the head offices of the mining houses in the last five years; this has spawned a number of smaller consultancy operations that now serve the mining companies as contractors.

Looking North: Reviving South African Mining's 'Historic Purpose'

It could be argued that the skills, technology and capital base developed from South Africa's mining industry would be the most direct source from which other African countries could have drawn these ingredients for industrial growth.[2] This process is now gradually unfolding, as economic liberalisation in many African countries re-opens foreign opportunities to South African mining companies, as well as those myriad firms providing mining equipment, technology and services. The involvement of companies like Anglo American in Zambia's copper industry is being rekindled following the company's acquisition of the Konkola copper mine.[3] Planned investments there (for refurbishment and expansion) are in the order of $800m., to give the mine at least a thirty-year life. The Konkola Deep project provides a useful indication of the importance of cross-border developments for the South African mining industry. Of the $520m. to be spent on expansions at Konkola, some $300m. is to be spent in South Africa on goods and services (mining equipment, consumables, design and engineering and management services, etc.). Further, of the $277m. to be spent on refurbishment, at least $158m. will be used to buy South African goods and services. Over the life of the mine, R13bn. will return to South Africa in payment for consumables and capital equipment (Menell, 2000).

Anglo American and other mining companies like Rand Mines had also accumulated mineral properties in various parts of Africa during the decades preceding isolation. In the 1990s, these were revisited and assessed more closely for commercial value. The scale of these projects has varied greatly. Rand Mines subsidiary Randgold Resources was formed to explore certain relatively small deposits in West Africa. Avmin (the base metals company spun off from Anglovaal) undertook exploration drilling in Zambia and gained a foothold through copper and cobalt processing; it is now developing a more substantial operation to recover cobalt from the slag left by previous decades of mining on the copperbelt. Anglo American is also moving back into Zambia, and taking the lion's share of state-owned mining company Zambia Consolidated Copper Mines (in which it had retained a minority - and more or less silent - interest during Zambia's years of nationalisation). Iscor Mining hoped to have secured a head start in the race for copper-cobalt mines in the Democratic Republic of Congo; the on-going war in that country, however, has prevented any commercial benefit being realised. Diamond giant De Beers had moved quickly to secure diamond interests in Angola; the resumption of the government's war with rebel movement Unita has put most organised mining activity there on hold.

Notwithstanding the South African interest in 'old' properties around the continent, it was the venture capital markets of countries like Canada that fed the renewed interest in African mineral exploration in the early to middle 1990s. Small ('junior') exploration companies made brave inroads into countries liberalising access to mineral opportunities. These companies secured mineral concessions and began exploring areas of mineral potential, many of which had been ignored for decades. Those that were successful in this high-risk activity would then usually sell or joint venture their discoveries with a large mining company, which would have the capital and expertise; in many instances, these companies have been South African.[4]

HIV/AIDS: Paying the Price for Migrancy

While the first cases of HIV-AIDS were encountered on South African mines in the 1980s, the industry has been generally slow to respond. The occurrence of HIV-AIDS in mining has both a historical-political and a future-economic significance. For the virus has feasted on the social conditions created by labour migrancy, which was initially evolved and institutionalised by the mines' insatiable demand for unskilled labour. The stress which migrancy placed on the fabric of traditional community and family relationships is a core factor in the vulnerability of mine workers to the spread of HIV-AIDS. While the practice of housing workers in single-sex hostels has come under criticism from unions and has been generally reviewed by mining companies, high levels of prostitution exist around these institutions and have helped spread the disease.

Recent studies by AngloGold showed HIV infection rates as high as 33 per cent at some of its South African operations (Swindells, 2001a); overall, about 23 per cent of AngloGold's employees are HIV-positive (Swindells, 2001b). Gold Fields has estimated that by 2006, its cost per gold bar produced could be increased by 10 per cent because of the disease's impact on its employees.

Figures quoted by the World Bank indicate that 30 per cent of gold miners and 20 per cent of coal miners in South Africa are HIV-positive, and that the impact of sickness and death among this group could raise mining company costs by a fifth. According to World Bank principal mining analyst Craig Andrews (2000):

> Because the disease is in its early gestation period and because we know it takes between five and seven years for a person to start dying of AIDS, that means that the full impact has yet to be felt on the mining industry in South Africa. In two, three, five years, fully 20 per cent or 30 per cent of the work-force in those mines will start to die. These are highly trained people, they are people in their prime productive years, between the ages of 25 and 45.

While the responses of both government and the mining companies to the impact of the virus have been hesitant and slow, there is a growing pool of data being developed by health planners and economists who are still scrambling to come up with a comprehensive response to the disease. It was reported in early 2001 that major South African mining companies had reached agreement with the National Union of Mineworkers to allow the performance of anonymous saliva tests for AIDS; the union had previously opposed the idea as a violation of workers' human rights (Swindells, 2001a). At the same time, Impala Platinum (the world's second largest platinum producer and employer of 28,000) announced that it had begun to assess the impact of AIDS on its operations (Swindells, 2001b).

Government Policy: In Search of the Greater Good

The African National Congress government has taken a cautious approach to instigating any major minerals-related issues. Despite the process of reviewing mineral policy being slow, the principles which government aims to pursue have been fairly clear. These were recently reiterated by the Minister of Minerals and Energy Phumzile Mlambo-Ngcuka (2001):

> Mineral resources are the common heritage of all South Africans; the state is the custodian of South Africa's mineral resources; the impact on the environment and sustainable development cannot be compromised; rights and developmental needs of workers and communities do matter; beneficiation and vertically integrated operations should be promoted through incentives; black economic empowerment should be promoted; government must provide accessible, efficient services to the industry; and the regulation of the diamond industry must be improved and streamlined.

There are, however, certain policy imperatives that it is pursuing; perhaps the most important relates to the 'ownership' of mineral rights. In essence, the aim is gradually to change the current system of private ownership of mineral rights to a system in which these rights will be vested in the state and be 'rented' to companies for certain finite periods of time. In Minister Mlambo-Ngcuka's words: 'We do not believe that black companies - and in some cases foreign companies - will find it easy to enter the South African economy if we do not improve the regulatory environment'. She was at pains, however, to dispel any fears of state over-involvement: 'the South African government in no way seeks direct participation in exploiting South Africa's minerals ... The actual development of mineral reserves, government believes, should be left to the private sector' (Mlambo-Ngcuka, 2001).

The argument against the current system (private ownership of mineral rights) is that indefinite control over a potentially mineable mineral deposit is not healthy for national economic development, and could effectively 'sterilise' mineral wealth if the company that owns it does not have the ability or the will to exploit it. This is especially relevant in South Africa, where mining companies have traditionally been large and powerful, controlling large inventories of minerals in the ground.

There are various interests which government aims to satisfy, as President Thabo Mbeki's speech to the Chamber of Mines reflects:

> ... as we seek to establish a new regime with regard to the issue of mineral rights, we will do this in a manner that seeks to secure the widest possible national consensus on this matter. I believe that this consensus should be based, among other things, on a common determination to redress the racist legacy we have all inherited; the need not to impact negatively on current mining, planned investment, property rights and the rule of law; the opening up of mining to new domestic and international entrants, including moving towards best international practice; and the further elevation of the mining industry as a central contributor to the fundamental objective of the overall reconstruction and development of our country and the continuous improvement of the quality of life of all our people (Mbeki, 2000).

Industry's counter-argument is that a high-risk and highly capital-intensive venture like mining requires developers to have long-term certainty about their right to the ground that they mine. Debate over the issue is ongoing and often vociferous (some observers have equated state custodianship of mineral rights with nationalisation of company assets). The fact that many mining countries successfully operate state-controlled systems of mineral rights is the best indication that policy will continue to move in this direction, despite logistical complications related to implementation and compensation.[5] Chamber of Mines president Rick Menell acknowledged the point clearly in his address to the Chamber's annual general meeting in November 2000:

Industry adoption of the transformation agenda is underpinned by the fundamental shift of attitude that has emerged on the issue of privately-owned mineral rights over the last two years. Consensus has moved from determined defence of privately-held rights to working through a process of embracing radical change in the mineral rights dispensation which will facilitate the transfer of all privately-held mineral rights - with the exception of current mining and development properties - to state ownership. This changed approach is based on the recognition that the State is, and must be, the custodian of the nation's mineral resources. These resources are a national asset and, in this sense, belong to all South Africans (Menell, 2000).

What the industry wants, however, is to retain the status quo on properties where mining or development is already underway.

Black Empowerment

The predominantly white complexion of ownership in mining companies has prompted government to put some pressure on the industry to make its management composition more representative of the population at large. Indeed, with a history so entwined with the evolution of migrant labour and rural reserves (later 'homelands'), the mining industry's image of a white bastion still stands almost unchallenged. However, there have been black economic empowerment initiatives to redress this, driven more often than not by the traditional mining companies themselves.

The early limelight on this issue was focused on JCI, which controlling shareholder Anglo American put on the market in 1994 after extracting its strategic platinum division (as Anglo Platinum), to facilitate black involvement in JCI's gold, coal, chrome and antimony assets. Various factors conspired to prevent this unbundling from becoming a success story, and 1997 saw the 'spectacular collapse' of the JCI empowerment structure.[6]

On a smaller scale, however, there have been more successful pioneers, such as companies like African Rainbow Minerals in the gold sector and Kuyasa Mining in the coal sector. African Rainbow Minerals, run by black businessman Patrice Motsepe, now successfully operates these previously marginal mines by applying a cost structure and methods more suited to the conditions. In 1997, Kuyasa purchased the Ikhwezi Colliery in Mpumalanga province, hailing it as the country's first mine to be wholly owned by a black empowerment group. The Commonwealth Development Corporation provided a loan of R15.2m. and Ingwe Coal (South Africa's largest coal producer) promoted the project and provided technical and marketing assistance.

In 1998, substantial coal assets were contributed by leading players Ingwe and Amcoal (now Anglocoal) towards the formation of empowerment company New Coal; as a result, this company would hold eight per cent of the country's coal reserves, produce 18m. tons of coal a year, and have its own export entitlement in Richards Bay Coal Terminal, South Africa's main coal export facility. New Coal's assets include the 12m.-ton-a-year Matla colliery (injected into New Coal by Ingwe and Amcoal, who owned 50 per cent each), the 1m.-ton-a-year Glisa colliery, the 1m.-ton-a-year New Clydesdale colliery and the 4m.-ton-a-year underground section of Amcoal's Arnot colliery. The export entitlement at RBCT was 800,000 tons a year.

Diamond giant De Beers has also put together arrangements to make available certain assets to an empowerment venture, and former premier of the Gauteng province, Tokyo Sexwale, has spearheaded the formation of his own company Mvelaphande Diamonds.

The Chamber of Mines of South Africa has itself recently begun a collaborative project with the Department of Minerals and Energy, to develop a supportive environment for new entrants to the industry. The so-called Bakubung Initiative includes the establishment of a fund of over R1bn. to finance the acquisition of mineral rights by previously disadvantaged South Africans. The words of the Chamber president (Menell, 2000) mean to leave no doubt about the industry's commitment to the transformation agenda:

> This is an agenda that enjoys the unqualified support of the industry, which acknowledges that many existing mineral rights were acquired at a time when the majority of South Africans were not allowed by law to obtain ownership of mineral rights. It is also accepted that Government has a ballot-given mandate to transform all segments of South African society, including pursuit of a sustainable economic transformation. Our mining industry's objective, through the establishment of a co-operative partnership with Government, and indeed the National Union of Mineworkers and other employee organizations, is actively to promote the transformation process.

In this context, the industry appears ready to give substantial support to transformation initiatives in return for guarantees that would safeguard mineral rights at existing mining and development operations.

In pursuit of black empowerment, some mining companies have also created buying policies that seek to source a level of equipment and services from companies owned by people from 'previously disadvantaged communities'. Anglo Coal, for instance, spent R134m. of supply transactions with 107 black empowerment ventures during the year 2000. The products purchased and services procured included conveyor belts, conveying accessories and idlers, computer equipment, cleaning services, petroleum and diesel supplies. In the same year, Anglo Platinum spent R106.8m. on procurement and business development from black empowerment companies and small- and medium-sized enterprises. Services procured have included slimes dam maintenance, underground cleaning, transport, catering and garment manufacturing.

Local Beneficiation: Claiming More of the Value Chain

It has always been a stated government mineral policy imperative that minerals should be the subject of greater beneficiation, in order to retrieve more value from the raw material (gold, for instance) before it is sold to outside markets. The mining industry was in the past not a particularly active participant in this endeavour, claiming that the extraction of gold metal from ore was in any event a process of considerable beneficiation.[7] In the last few years, however, a change of heart has been evident - due in large part to the growing commoditisation of gold and consequent efforts to further brand and market gold as jewellery.

Gold producer AngloGold has taken a particular lead in embracing the need to bolster the market for gold. The company's African roots have become a central theme in its corporate branding, and this process has extended to its support for the development of local jewellery design talent, and the establishment of a museum to raise public awareness and appreciation of historic African traditions in crafting gold.

The high level of common interest in this particular sphere is reflected by a recent tripartite delegation led by the minerals department's deputy minister Susan Shabungu to visit three West African countries (Ghana, Mali and Senegal) to examine gold jewellery fabrication techniques. The delegation was tasked to investigate the feasibility of transferring jewellery-manufacturing skills to South Africa to provide jobs, in the first instance, for retrenched gold-mining industry employees. All three countries have centuries-old traditions of producing gold jewellery that is both attractive and affordable to their own people. The delegation was reportedly convinced that South Africa could emulate these industries in some form:

> It has been agreed - although the concept requires a great deal more work - that with the assistance of West African governments, their countries' gold producing sectors and master goldsmiths, development of a similar industry in South Africa can be achieved and that the initiative could form the bedrock of enhanced beneficiation and a rural-based economic growth sector providing much needed employment opportunities (Menell, 2000).

A similar process is underway in the platinum industry, where producers have been part of a recent jewellery design and manufacturing project. This project, in partnership with Technikon Pretoria (a tertiary educational institution similar to the British polytechnic), provides the opportunity for students and jewellers to specialise in the techniques of platinum jewellery design and manufacture, to encourage an involvement in the growing jewellery industry.

Conclusion

As South Africa finds its way out of its isolated apartheid past, the mining industry is being called to account on a number of levels - by government, shareholders and civil society. On a political level, it will be required to demonstrate further its commitment to including in its ranks a greater number (and higher level) of people from previously disadvantaged groups. On a business level, the demand for globally competitive returns will continue to exact more efficiency. And society generally will be looking to local mining companies to polish their image as responsible corporate citizens in the community and natural environment.

Geographically, the industry will continue to be drawn into the global sphere through acquisitions, mergers, foreign mining investments and the supply of goods and services. The impact will be felt in growing levels of conformity in corporate governance, best technical practice, labour relations and environmental responsibility. The resulting pressures will often be contradictory, and the success with which it can balance these will in large part determine the nature and extent of its role in building South Africa's economic and political future.

Notes

1. Although unskilled wages had been rising during the 1970s, an unskilled worker on a gold mine could expect to be paid on average about 15 per cent of a skilled worker's wage.
2. 'The truncated access of South African firms to international capital markets and investment opportunities abroad precluded mining houses from pursuing the most natural path of expansion. This path would have been to do for the rest of the developing world what they had done for South Africa: be a conduit of capital from the international financial centres to mining opportunities in high-risk developing environments' (Malherbe, 2001, p. 70).
3. Although Anglo American remained a silent shareholder in Zambia Consolidated Copper Mines during the years following the effective nationalisation of the copper industry, it had relinquished any operational role.

4. Among the more well known of these ventures has been Sadiola Mine in Mali; the concession was secured and explored by Canadian junior company Iamgold, which then formed a joint venture with AngloGold to develop and operate the mine.

5. The Chamber of Mines is also concerned about the level of the Minister's discretion in the bill, and are keen to see the state 'set objective criteria for the granting of mineral rights with continuity of tenure from prospecting through life-of-mine operations, with the minimum of administrative and political latitude and with right of appeal to an independent judiciary. These benefits are as important to new entrants to the industry as to existing participants' (Chamber of Mines, 2000b).

6. Among the undermining events at the time were the cyclical fall in commodity prices together with the widely publicised crisis in gold mining, making the industry less attractive to black investors. Also, poor equity performance on the Johannesburg Stock Exchange in the aftermath of the emerging markets crisis undermined the classic empowerment structure, the success of which depended on good share price performance (Malherbe, 2001, p.82).

7. The term 'beneficiation' could be defined more narrowly, as it is by the Department of Minerals and Energy's Mineral Economics Directorate (formerly the Minerals Bureau) to denote 'the upgrading of a primary ore to a stage where the component or component products can be used in the manufacturing process' (Department of Minerals, 2001, p. 14). But in the public domain, the term is used more broadly to indicate areas where the metals produced by SA mines could be turned into a final, saleable product in the retail marketplace.

References

Andrews, C. (2000), Address to Mining 2000 Conference, Melbourne, Australia. Reported in the *West Australian* newspaper, 23 September 2000.

Chamber of Mines of South Africa (2000a), *Annual Report 1999.*

Chamber of Mines of South Africa (2000b), Media Release: Background Information Provided by the Chamber of Mines on the Minerals Development Bill, 18 December 2000.

Department of Minerals and Energy (2001), *South Africa's Mineral Industry 1999-2000*, Mineral Economics Directorate (Minerals Bureau), Pretoria.

Hobart Houghton, D. (1967), *The South African Economy*, Oxford University Press, Cape Town.

Jones, A. (2000), 'Further South African Gold Industry Transformation Likely in 2000', *African Mining*, January-February 2000, Resource Publications, Johannesburg.

Malherbe, Stephan (2001*), A Perspective on the South African Mining Industry in the 21st Century*, Graduate School of Business, University of Cape Town.

Mbeki, President Thabo (2000), Opening Address to the 110th Annual General Meeting of the Chamber of Mines of South Africa, 7 November 2000.

Menell, R. P. (2000), Presidential Address to the 110th Annual General Meeting of the Chamber of Mines of South Africa, 7 November 2000.

Mlambo-Ngcuka, The Hon Phumzile (2001), 'Government's Role in the Development of South Africa's Mineral Resource', presentation at the Investing in African Mining Conference/Indaba, February 2001, Cape Town.

Nattrass, N. (1995), 'The Crisis in South African Gold Mining', *World Development*, vol. 23, pp. 857-68.

Swindells, S. (2001a), 'South African Mining Firms Test Workers for AIDS', article for new agency Reuters, 13 February 2001.

Swindells, S. (2001b), 'South African Miner Implats to Begin AIDS Impact Study', article for news agency Reuters, 8 February 2001.

4 South Africa's Manufacturing Economy: Problems and Performance

ETIENNE NEL

Introduction

The manufacturing sector is a vitally important component of the South African economy even though that economy is gradually assuming post-industrial characteristics, with the tertiary sector now the largest employer and economic contributor. The manufacturing sector currently generates the second biggest share of the GDP, namely 23.9 per cent (Editors Inc, 2000). It would be fair to argue that one of the key reasons why South Africa is the largest economic power on the continent is the sustained growth, over a century, of manufacturing, and the long-established interdependence which developed from the late nineteenth century between the mining, energy and manufacturing sectors. This relationship created the key catalyst in South Africa's economic ascendancy, namely what Fine and Rustomjee (1996) have termed the Mineral-Energy Complex (MEC). But as South Africa enters the twenty-first century, all is not well with the manufacturing sector. Its problems occur at a time when economic growth in all sectors is vitally needed to address desperately high levels of poverty which seem to have been exacerbated, not diminished, in post-apartheid South Africa (Lester et al., 2000).

The lifting of sanctions did not prove sufficient to revitalise manufacturing growth as many anticipated (Blumenfeld, 1994). Structural weaknesses, low productivity growth levels, low skill levels, cheap imports, labour unrest and difficulties encountered in attempting to become internationally competitive after years of protection under import substitution policies, have all taken their toll (EIU, 1998; Editors Inc, 2000). The seriousness of the situation facing the sector has been alluded to by Bill Cooper, President of the Steel and Engineering Industries Federation of South Africa, who stated, 'the greatest challenge facing the manufacturing industry as it starts the new millennium is its survival' (Cooper, 2000, p. 24).

The above realities are of serious concern to the country's future socio-economic well-being. Rogerson (1995, p. 185) notes that 'most policy observers see the reinvigoration of South Africa's stagnant manufacturing sector as the key sectoral issue in long-term economic development'. In parallel, the national government-labour-business debating forum, Nedlac, sees economic growth and employment creation as the central challenge facing the country (Nedlac, 2000). Whilst the government has strenuously sought to encourage the international competitiveness of the economy and the growth of new businesses, the basic reality, as argued by Cassim (1988, p. 14), prior to the demise of apartheid, is that 'any new government will not be at liberty to fabricate an economy. It will in fact inherit an economy which imposes a certain logic and rigidity on the course of future development'. In addition, the international division of labour imposes its own constraints on the ability of the country to participate adequately in the global system, particularly in the light of the poor competitiveness of many firms (Editors Inc., 2000). In this chapter the evolving history of the country's manufacturing economy is outlined before switching attention to an examination of recent debates and trends in the sector. Government policy is then outlined before concluding with an assessment of the status quo and future prospects.

The Development of the Manufacturing Sector

It was the mineral discoveries in the late-nineteenth century which provided the justification and requirement for the emergence of a modern, capital-intensive manufacturing sector capable of meeting the increasingly sophisticated needs of the mineral sector and the growing population (Coleman, 1993). Sustained growth continued through the first half of the twentieth century and was significantly boosted by post-World War One and Two booms. A key factor promoting these booms was the introduction from the 1920s of import substitution policies which significantly aided the expansion of the consumer goods industry (chapter 16). In parallel, major government investment in monolithic state-supported enterprises such as SASOL (the oil corporation) and ISCOR (the iron and steel corporation) injected significant capital into the manufacturing economy, inducing noteworthy cumulative growth effects in the country's space economy.

By the 1950s near self-sufficiency in the capital and consumer goods sectors had been achieved (Nel, 2000). However, rapid growth had been at the expense of the black majority who were subjected to institutionalised racial discrimination and exploitation through a system dubbed as 'racial Fordism' (Rogerson, 1991). Historically, as a result of the above considerations, production and demand have been focused on the internal markets and the needs of the white, affluent minority, with labour allocation and control being subsumed within racist ideology (Cassim, 1988; Rogerson, 1991) and little emphasis being placed on exports till the finding of the Reynders Commission in 1972 (Nel, 2000).

Whilst economic growth, based on these biases, could be sustained for a period of time, by the 1980s the imposition of sanctions, over-reliance on the mineral sector, decades of exploitation, skill and capital shortages and gradual weakening of the mineral economy all began to take their toll (Cassim, 1988). In addition, many industries were no longer operating at optimum levels as a result of decades of being sheltered behind tariff barriers, leading Rogerson (1991, p. 354) to refer to the 'miserable performance of South African industry' as a distinctive feature of the economy. In addition, emerging crises brought into question the nature of the whole relationship between capitalism and apartheid and the reproduction of apartheid itself (Cassim, 1988), such that by the 1970s the fundamental basis of the manufacturing economy had started to deteriorate (Bell, 1995; Bell and Madula, 2001). Although by 1970 South Africa had achieved a relatively advanced and diversified manufacturing sector thereafter 'manufacturing output has stagnated and employment declined' (Bell and Madula, 2001, p. 4). Factors which have led to manufacturing stagnation include: declining gold exports and gold prices, the reduction in global commodity demand from the early 1980s, the global debt crisis, decline in the value of the rand, reduced levels of internal investment, foreign exchange shortages and skill shortages (ibid., 2001).

By the 1990s very real job losses started to occur. Between 1990 and 1994, the manufacturing sector lost 140,000 jobs (DTI, 1998) and by the end of the decade evidence of deindustrialisation was starting to emerge in once prosperous manufacturing belts, such as the KwaZulu-Natal Midlands (Nel and Hill, 2001). Deindustrialisation needs to be seen as both a response to inherent weaknesses and internal market changes as well as the effects of re-integration into the global economy in the 1990s. One of the causes of job loss and closure has also been the suspension of apartheid government support for industrial establishments in the former black homelands. By the 1990s low labour absorption capacity and rising unemployment in the manufacturing sector and in the economy generally had become issues of serious concern. In the view of Loots (1998, p. 319) 'one of the biggest problems facing South Africa today is the high level of unemployment. It seems that economic growth, generally regarded as the creator of employment, is not able to create sufficient employment opportunities for the growing South African labour force'. Unemployment rose from 6.7 per cent in 1960 to 33.9 per cent in 1996 (StatsSA, 2000), whilst the labour absorption rate declined from 79.6 per cent to effectively zero (Loots, 1998).

The commonly held view, as espoused in the Industrial Strategy Project, is that low productivity growth and the entrenched import substitution manufacturing economy are the key cause for the slow-down in the manufacturing economy. Bell (1995) takes issue with these findings and ascribes this trend to negative macro factors such as the series of foreign exchange shocks which the country experienced and the negative effects of import liberalisation. In reality, one is witnessing the impact of a combination of sector-specific and economy-wide changes and crises and ascribing them to a single explanation would seem to be unrealistic. The end product is that the manufacturing economy is clearly transforming and that the long-term future is uncertain. Rogerson (1995) identifies three key themes which affect the future growth trajectory of manufacturing in South Africa: the need to shift from an import-substituting to an export economic focus, the role of foreign direct investment and multi-national corporations and the potential for flexible specialisation to create new growth opportunities.

The Minerals-Energy Complex Debate

The nature of South Africa's economic evolution has had a defined impact on the current structure and nature of the manufacturing economy. In this section attention shifts to consider one of the key outcomes of these historical processes. Arguably the key academic debate in the last decade regarding South Africa's manufacturing sector revolves around the concept of the Mineral-Energy Complex (MEC) as postulated by Fine and Rustomjee (1996). In addition to catalysing critical debate regarding the structure of the manufacturing economy, it has also drawn attention to the fact that manufacturing has been skewed in terms of the over-representation of the mineral- and energy-based sectors. According to Fine and Rustomjee (1996, p. 5), 'the central theme of our argument is that what will be termed the Minerals-Energy Complex (MEC) lies at the core of the South African economy, not only by virtue of its weight in economic activity but also through its determining role throughout the rest of the economy'. By the mid-1980s the MEC was contributing over one-quarter of GDP. Moreover, the MEC is not just a core set of industries and supporting firms. It is also bolstered by an historically entrenched system of institutions, financial support structures and labour control, creating a defined system of accumulation which 'has had a pervasive effect on almost every aspect of the economy' (ibid., p. 14).

The core focus of the MEC is a small number of giant, state-supported firms which have dominated the national economy for nearly a century. State policy was married to emerging Afrikaner capitalist power for much of the twentieth century. According to Fine (1997) this created a series of major corporations which have come to dominate the electricity, iron and steel, transport, oil, chemical and arms sectors. He argues that the MEC's growth retarded the development of other sectors and industrial diversification in general. In addition, Fine and Rustomjee (1996) argue that the growth of manufacturing around primary production has been more important than import-substitution industry. They credit see the MEC with having significantly advanced the development of productive and infrastructural capacities around the core sectors of the economy. On the downside there has been a failure of vertical integration forward from the MEC into the rest of the economy (ibid.).

The MEC argument is contested by Bell and Farrell (1997) who argue that although the mining and energy are undoubtedly key sectors in the national economy, there is no evidence that the MEC, as a system of accumulation, has increased the dependence of the overall economy on these sectors. In addition, they state that South Africa's economic history did follow a typical import substitution trajectory from consumption goods to intermediate then capital goods and, further, that there is no evidence that the MEC's existence retarded industrial diversification or industrialisation in general. In their view the MEC is not undesirable nor did it cause dependence, instead its existence was an 'important and necessary step towards full-scale industrialisation' (Bell and Farrell, 1997, p. 610). In response Fine and Rustomjee (1998) point out that fundamental links exist between the MEC and other sectors of the economy, giving the appearance of diversification. They also point out that the MEC is making strides to globalise in order to maintain its pre-eminent position.

Broad Trends in the Manufacturing Economy

In this section attention shifts from a discussion of historical developments and academic debate to consider the broad features and trends which currently characterise the manufacturing sector. As suggested above, the recent history of the manufacturing sector has been disappointing with the total number of manufacturing employees declining from 1.4m. in 1981 to 1.29m. in 2000. Whilst overall employment remained static until the 1990s when decline set in, the overall number of firms appears to have increased quite significantly. However, it is apparent that the emerging firms are in the capital-intensive sectors and they are not compensating for the loss, in employment terms, experienced by labour-intensive sectors vulnerable to globalisation. Between 1991 and 1996 the number of primary activity manufacturing jobs decreased from 280,000 to 239,000, due mainly to rationalisation in the steel industry (DTI, 1998).

Spatially, it appears that the historical dominance of the core metropolitan areas of the country as the primary nodes of manufacturing activity remains largely unaltered as has been the case for the last century (Coleman, 1983). In terms of industrial census data, it would appear that in 1996 nearly 21,000 of the South Africa's 25,402 factories were located in the Gauteng heartland (Johannesburg, Pretoria and surrounding towns) or the Durban and Cape Town secondary cores and their hinterlands. In 1991 these areas had nearly 1.1m. of the 1.5m. manufacturing jobs and generated some R147bn. of the R191bn. turnover (CSS, 1996). Gauteng's manufacturing industry continues to play a critical economic role as 'one of the key economic driving forces of the South African space economy' (Rogerson, 2000a, p. 311) despite emerging constraints which are discussed below. A factor abetting the continued dominance of the core areas has been the demise of state regional development assistance, which impacted severely on the peripheral eras previously favoured under state decentralisation assistance policies (Rogerson, 1991). As a result, 'the new industrial geographies of post-apartheid South Africa appear set to be dominated by a re-focussing of manufacturing activity around the large metropolitan centres, (and) the demise of the industrial base of several favoured decentralised growth points' (ibid., p. 364).

Despite the continued dominance of the core areas, selective crises are also occurring within the cores, such as the closure of textile firms in Cape Town, whilst the Witwatersrand area of Gauteng is clearly experiencing a contraction in its manufacturing economy. Between 1980 and 1991 there occurred a structural economic transformation in this area as it experienced a 39.5 per cent fall in industrial employment due to the demise of labour-intensive textile and clothing production, a fall in mining output and rising labour costs. This downward trend has extended through into the 1990s, with the number of establishments falling by nearly 200 and a corresponding loss of nearly 100,000 manufacturing jobs in Gauteng (Rogerson, 2000a). Whilst broad overall decline is apparent, global statistics mask the rapid decline of the central Johannesburg region and the not insignificant new growth which is taking place both in the Midrand and Pretoria sub-regions of the province and in the small, medium and micro-enterprise sector (ibid.).

One of the key themes influencing manufacturing has clearly been the impact of trade liberalisation. There are strong policy arguments in favour of trade liberalisation and the removal of import and export barriers in line with the requirements of the WTO, but as Bell (in Zarenda, 1991) and Zarenda (1991) have argued, freer access to international trade will inevitably harm the weaker, less competitive sectors. Despite this, there has been an increase in manufactured exports as a result of policy and economic shifts, such that between 1989 and 1997 manufactured goods increased from 20 to 30 per cent of total exports (EIU, 1998). On the negative side, the vulnerable consumer goods industries have experienced closure and job loss. For example in 1998, 20,000 clothing and textile workers lost their jobs. Despite the growth in exports which has taken place, Cooper (2000, p. 24) notes that 'South African industries are finding difficulty both in penetrating new export markets and in coping with international competition in domestic markets'.

High and sustained levels of foreign investment (chapter 16) are clearly critical to ensuring the growth of the manufacturing sector although very low levels of domestic investment in productive capital have been a hallmark of recent decades (Lester et al., 2000). In terms of Foreign Direct Investment (FDI), funding is constrained by the general global slowdown in investment in developing countries, political uncertainties in South Africa, decreased south-east Asian investment, economic instability and currency volatility in South Africa (Heese, 2000). Sectors which have benefited from FDI are IT and telecommunications, energy and oil, food and vehicles. Between 1994 and 1998 250 foreign companies invested in the country (DTI, 1998), but Rogerson (1995) stresses that there are limitations to the volume of FDI and multi-national corporate investment which can be expected in post-apartheid South Africa.

A number of other factors which are increasingly impacting on the nature and operation of the South African manufacturing economy in general and the Gauteng area in particular have been identified by Bell and Madula (2001), the Department of Trade and Industry (DTI) (2001) and Rogerson (2000a). These include:

- the gradual shift of the economy towards service-based economic sectors, including more knowledge-intensive economic activities, and the related need to focus attention on innovative economic activities;
- a decline in the number of employment opportunities in manufacturing as a result of falling commodity prices, skills shortages, the poor performance of gold and the global economy in general;

- increasing informalisation of the economy and the growth of the small business sector;
- new growth areas such as Midrand and the impact of new planned development corridors and special Industrial Development Zones.

Government Policy and Spatial Support Mechanisms

The government supports the manufacturing sector in policy and practical terms and in this section key policy and strategy issues are discussed, including the defence programme. The post-apartheid government, through the Department of Trade and Industry (DTI), has actively sought to promote economic development, primarily through the encouragement of export-orientated industrialisation, the search for investment, support for small business development, the encouragement of defined manufacturing clusters and new spatial development strategies, as detailed below. The DTI's industrial policy has the following objectives:

- to consolidate competitive advantages that upstream industries and primary manufacturers enjoy as a result of past policies which favoured the capital- and energy-intensive manufacturing sectors;
- to transfer these advantages to labour-intensive sub-sectors of manufacturing, many of which supply inputs into primary manufacturing (DTI, 1998).

Key interventions include the provision of export marketing and investment assistance and the establishment of an Export Credit and Foreign Investment Reinsurance Scheme (RSA, 2000). Clear encouragement has been provided to various key clusters such as vehicle manufacturing and chemical production, whilst incentives have been provided to potential foreign investors.

Overall government policy is ultimately framed within the context of broad national economic policy as encapsulated in GEAR (the Growth, Employment and Redistribution strategy). This commits the country to the pursuit of broadly neo-liberal strategies which seek to encourage the free market, promote greater participation in the global economy and attract investment. In terms of the pursuit of GEAR, the DTI, through its industrial policy, 'strives to achieve a balance between greater openness and improvement in local competitiveness whilst pursuing a process of industrial restructuring which aims at expanding employment opportunities and productive capacity' (RSA, 2000, p. 17). Regrettably, GEAR has not led to the anticipated levels of employment creation. By 1998, 18 months after its introduction, 80,000 jobs had been lost in the economy, whilst GEAR had predicted the creation of roughly 200,000 jobs (Lester et al., 2000). Reasons for this include international economic volatility, lower than expected levels of investment and, according to Fine, (1997, p. 135), the fact that 'there is no unity of vision of what constitutes industrial policy and no unity of purpose in formulating and implementing such policy'. Whilst GEAR clearly accords with the current global economic orthodoxy and has the support of the global financial bodies it has angered the unions and workers displaced by the poor competitiveness of numerous industrial sectors.

A key thrust of the post-apartheid government has been targeted support for the small business or Small Medium and Micro Enterprise (SMME) sector. Such support is justified on the grounds that 44 per cent of new market entrants are absorbed by this sector and it is logically seen as a strategy which can be used to empower previously disadvantaged communities. In pursuit of this ideal, the DTI established the Ntsika Enterprise Promotion Agency actively to support SMME developments and to establish strategically sited Local Business Service Centres (LBSCs) to provide them with support and advice. In parallel, an institution known as Khula was established to assist emerging firms financially (Nel, 2000). Despite the good intentions of this strategy, it is unfortunate that only a small percentage of new micro-enterprises graduate to become small enterprises with over ten workers (Rogerson, 2000b). By 2001 the government was beginning to admit that the sector was experiencing difficulties. The Minister of Trade and Industry, Alec Erwin, reported that 117,426 state-supported small businesses had failed between 1997 and 2000, costing the country R68m. Reasons for their failure include weak entrepreneurial culture and the shortage of technical and management skills.

One of the most controversial issues regarding the manufacturing sector and government participation within it relates to the R29.9bn. defence deal approved in 1999. The government claimed that in return for re-arming itself from foreign suppliers the industrial offset arrangements with the supplying countries would lead to R104bn. of return investment in the country, creating 65,000 jobs (Batchelor and Dunne, 2000). The scheme has not been without its controversy and allegations of corruption. In addition, some analysts regard the award of contracts to foreign firms as a serious blow to the domestic arms industry. Historically, the development of the South African arms industry 'distorted the trajectory of the country's economic development and imposed a number of long-term economic costs on the economy' (ibid., p. 425). It absorbed scarce capital and labour resources and crowded out non-military investment and research and development. Now, when some of that capacity could have been used, it has effectively been disregarded. Concerns are also being expressed about whether the anticipated counter-investments will be as significant as was initially anticipated. Although there has been investment in sub-sectors such as those supplying specialised aircraft components, major investments, such as the proposed stainless steel plant at Coega (close to Port Elizabeth), appear to be in abeyance.

A new, draft Department of Trade and Industry (DTI) policy document released in 2001 indicates how government will seek to promote economic development in future (DTI, 2001). The document argues that whilst reliance on the free market will continue to be a key element of government strategy, there is also a need to address market failure through providing capacities - both services and products - which are in short supply. The key elements identified as needing to be supported and encouraged are the promotion of innovation, integrated industrial development and, most importantly, support for knowledge-intensive manufacturing (chapter 7). As a result, 'the provision and enhancement of knowledge will be increasingly the hallmark of our intervention on the supply side' (DTI, 2001, p. 8). Key objectives for the DTI include the promotion of SMMEs, black economic empowerment, employment creation and knowledge-based development.

Spatial Policy

Prior to 1991 the South African government pursued a very deliberate and well-funded regional development strategy designed to support the apartheid-created homelands. In that year the policy was terminated and, for a brief period, replaced with an 'aspatial' industrial support policy. In 1997 this was superseded by a 'tax holiday scheme' to encourage industrial development throughout the country. This too is in abeyance and policy has swung back to spatially focussed considerations in recent years (Nel, 2000).

The key instruments in this respect are the Spatial Development Initiatives (SDIs) and the Industrial Development Zones (IDZs). The SDIs are supported by both the Ministry of Trade and Industry and the Ministry of Transport which seek to encourage investment, manufacturing and other economic activities along a series of defined transport corridors. The IDZs are state-of-the-art industrial zones with a defined export focus. The latter obviously share similarities with the export-processing zone concept which has met with mixed success internationally (Nel, 2000). Whilst considerable effort has been put into the promotion of SDIs, they ultimately rely on the commitment of the private sector, which with the major exception of the Maputo development corridor, does not yet appear to have brought about dramatic economic transformation in the areas which they transect. The Maputo corridor runs from Gauteng to the port of Maputo in neighbouring Mozambique and is based on the N4 Highway (Rogerson, 2001). Despite their limited success to date, as Rogerson (2001, p. 324) argues, SDIs are becoming 'a critical feature in the planning for reconstruction in post-apartheid South and (Southern) Africa'. Support for manufacturing, tourism, local economic development and infrastructure is a hallmark of the programme in the case of the Maputo corridor. Related to the SDI strategy is the continued promotion of resource-based industrialisation, a traditional core focus of the South African economy (Walker, 2001).

The establishment of IDZs has been slow. The Coega IDZ in Port Elizabeth is to be based on a deep-water port that has yet to be built. This IDZ has not been without its share of controversy and in 2001 doubts were cast on the scheme as result of reports of possible wrong-doing and the absence of a major anchor tenant. The other IDZs are all planned to be sited at major ports or airports. It was only in 2001 that the government announced incentives, valued at R3bn., which would apply in IDZs. These include R600m. for wage incentives to boost job creation and a 50-100 per cent investment allowance for investors undertaking approved projects. Concerns regarding IDZs include their high set-up costs, fears of the relocation of firms from other areas and even neighbouring countries and the possible 'leakage' of goods into the domestic economy.

Assessment and Future Prospects

As the above discussion reveals, South Africa's manufacturing sector has come through very challenging times. International competition, low levels of investment and domestic uncertainties have all taken their toll. Government clearly seeks to support manufacturing development, but policies such as IDZs appear to be slow off the ground, have not lived up to expectation in the case of the SMME sector or, in the case of the defence deal, have experienced considerable controversy and have not yet delivered significant tangible results.

The manufacturing economy does have numerous strengths including the well-developed nature of the MEC, good infrastructure and a strong financial system (Fine, 1997), all of which provide a basis for expansion. Key challenges include the need to increase overall productivity, employment and investment levels, but this is hindered by the very real loss of skills through emigration and the fact many of the key exports are not the product of labour intensive activity.

Invariably, as Cassim (1988) argued, it will prove difficult for South Africa to follow a radically different growth strategy from its historical precedent. In this regard it is in some ways quite remarkable that, despite the obvious current focus on SMMEs, past policies of support for the MEC have continued with little break from the past. The promotion of resource-based industrial strategies (Walker, 2001), most notably through a series of mega-projects, such as the Alusaf aluminium refinery at Richards Bay and Columbus Steel at Witbank (which combined equalled half of all manufacturing investment in the 1993-95 period) are cases in point (Fine, 1997). As a result, one strategic option may be to strengthen MEC-type firms with their established bases and potential, since productivity, employment and export potential are probably easier to encourage in areas where comparative advantage already exists rather than in new sectors that have yet to prove their potential.

Improving productivity and export production are critical to the growth of manufacturing capacity, although van Rensburg and Naude (1999, p. 269) show that there is 'little evidence of a statistically significant relationship between export growth and productivity growth' except in the chemicals and wood processing sectors. The manufacturing sector is clearly at a difficult juncture in its history, and as Loots (1998, p. 335) argues, 'new approaches are needed to expand employment in order for the broader population to participate in growth and benefit from it'. These will require a focus on established strengths and the development of new opportunities both within and outside of the manufacturing sector if overall goals of employment creation and economic growth are to be achieved. A new focus on 'knowledge-based' activity, as advocated by the DTI (2001) in its latest strategy document, may offer one such opportunity (chapter 7). There are, however, no recipes for instant success: the obstacles to significantly faster growth of South Africa's manufacturing exports are formidable in today's global conditions, and the significance of the southern African regional market (chapter 15) is likely to remain critical.

References

Batchelor, P. and Dunne, P. (2000), 'Industrial Participation, Investment and Growth: the Case of South Africa's Defence related Industry', *Development Southern Africa*, vol. 17, pp. 417-35.

Bell, T. (1995), 'Improving Manufacturing Performance in South Africa: a Contrary View', *Transformation*, vol. 28, pp. 1-31.

Bell, T. and Farrell, G. (1997), 'The Minerals-Energy Complex and South African Industrialisation', *Development Southern Africa*, vol. 14, pp. 591-613.

Bell, T. and Madula, N. (2001), *Where has all the Growth Gone? South African Manufacturing Industry 1970-2000*, Trade and Industrial Policy Strategies Annual Forum, Muldersdrift.

Blumenfeld, J. (1994), 'Sustainable Growth after Sanctions: Opportunities and Constraints', in Cole, K. (ed), *Sustainable Development for a Democratic South Africa*, Earthscan, London, pp. 1-27.

Cassim, F. (1988), 'Growth, Crisis and Change in the South African Economy', in Suckling, G. and White, L. (eds), *After Apartheid: Renewal of the South African Economy*, James Currey, London, pp. 1-15.

Coleman, F. (ed) (1983), *Economic History of South Africa*, HAUM, Pretoria.

Cooper, B. (2000), *Management Today: Yearbook 2000*, vol.15, no. 10, pp. 24-26.

CSS (Central Statistical Service) (1996), *Census of Manufacturing 1991, Report No. 30-01-02*, Central Statistical Service, Pretoria.

DTI (Department of Trade and Industry) (1998), *Annual Report: RP 57*, DTI, Pretoria.

DTI (2001), *Discussion Document: Driving Competitiveness: An Integrated Industrial Strategy for Sustainable Employment and Growth*, DTI, Pretoria.

Editors Inc. (2000), *SA 2000-01: South Africa at a Glance*, Editors Inc, Craighall.

EIU (Economist Intelligence Unit) (1998), *EIU Country Profile: South Africa 1998-1999*, EIU, London.

Fine, B. (1997) Industrial and Energy Policy, in J. Michie and V. Padayachee, (eds), *The Political Economy of South Africa's Transition: Policy Perspectives in the Late 1990s*, Dryden Press, London, pp. 125-149.

Fine, B. and Rustomjee, Z. (1996), *The Political Economy of South Africa*, Hurst, London.

Fine, B. and Rustomjee, Z. (1998), 'Debating the South African Minerals-Energy Complex: a Response to Bell and Farrell', *Development Southern Africa*, vol. 15, pp. 689-701.

Heese, K. (2000), 'Foreign Direct Investment in South Africa (1994-9) - Confronting Globalisation', *Development Southern Africa*, vol. 17, pp. 389-400.

Lester, A., Binns, T. and Nel, E. (2000), *South Africa: Past Present and Future*, Pearson, Harlow.

Loots, E. (1998), 'Job Creation and Economic Growth', *South African Journal of Economics*, vol. 66, pp. 319-36.

Nedlac (2000), *Annual Report on Social and Economic Conditions in South Africa*, Nedlac, Johannesburg.

Nel, E.L. (2000), 'Economies and Economic Development', in R. Fox and K. Rowntree (eds), *The Geography of South Africa in a Changing World*, Oxford University Press, Oxford, pp. 114-37.

Nel, E.L. and Wood, T. (2001), 'De-industrialisation and Local Economic Development Alternatives in Kwazulu-Natal, South Africa', *Geography*, vol. 86, 356-9.

RSA (Republic of South Africa) (2000), South Africa Yearbook: 2000-2001, http://www.gov.za/yearbook/economy.htm 1/3/2001.

Rogerson, C.M. (1991), 'Beyond Racial Fordism: Restucturing Industry in the "New" South Africa', *Tijdschrift voor Economische en Sociale Geografie*, vol. 82, pp. 355-66.

Rogerson, C.M. (1995), 'The Employment Challenge in a Democratic South Africa', in A. Lemon (ed), *The Geography of Change in South Africa*, John Wiley, Chichester, pp. 169-94.

Rogerson, C.M. (2000a), 'Manufacturing Change in Gauteng 1989-1999', *Urban Forum*, vol. 11, pp. 311-340.

Rogerson, C.M. (2000b), 'Successful SMEs in South Africa: the Case of Clothing Producers in the Witwatersrand', *Development Southern Africa*, vol. 17, pp. 687-716.

Rogerson, C.M (2001), 'Spatial Development Initiatives in Southern Africa: the Maputo Development Corridor', *Tijdschrift voor Economische en Sociale Geografie*, vol. 92, 3, pp. 324-46.

StatsSA (1998), *Census of Manufacturing: Statistical Release: P 3001*, StatsSA, Pretoria.

StatsSA (2000), *Stats in Brief: 2000*, StatsSA, Pretoria.

Van Rensburg, L. and Naude, W. (1999), 'Productivity and Export Growth in the South African Manufacturing Sector', *South African Journal of Economic Management Science*, vol. 2, pp. 269-89.

Walker, M. (2001) 'Resource-based Industrialisation Strategies: a Comparative Analysis of the South African and International Experience', *South African Geographical Journal*, vol. 83, 93-104.

Zarenda, H. (1991) 'Directing Industrial Strategy in South Africa: Policy Choices for the Nineties', *Development Southern Africa*, vol. 8, pp. 387-94.

5 Tourism – A New Economic Driver for South Africa

CHRISTIAN M. ROGERSON

Introduction

On 9 February 2001 President Mbeki in the State of the Nation address at the opening of parliament announced an integrated action plan aimed at accelerating economic growth and development. The plan represents a cluster of linked action plans focused on specific sectors and aims to enhance the competitiveness of the national economy as well as stimulate high-growth areas. Among the several 'high growth areas' that were identified for priority intervention is tourism (Dobson, 2001). Since 1994 tourism has been viewed as an increasingly important sector of the economy with the potential to assume a critical role in achieving the goals and objectives for reconstruction and development (The Cluster Consortium, 1999a). At the 1998 Job Summit tourism was recognised as the sector 'having the greatest potential to reduce unemployment in the country' (Mahony and van Zyl, 2001, p. 4). According to the recent investigation by The Cluster Consortium (1999b, p. 17), 'South Africa has the potential to become a world-class player in tourism and to generate significant employment and economic opportunities throughout the country'. The World Travel and Tourism Council (1998) asserts that tourism has the potential to be one of the country's most important drivers of economic growth in the 21st century.

By 1998 it was estimated that tourism employed approximately 740,000 people (1 in every 16 workers) and contributed 8.2 per cent to South Africa's Gross Domestic Product (Elliffe, 1999, p. 4; Department of Environmental Affairs and Tourism [DEAT], 2000a, p. 2). In terms of foreign exchange earnings tourism is ranked as the third largest earner after manufacturing and mining with several analysts suggesting that it is poised to overtake mining in the near future (DEAT, 2001). Within the global tourism economy South Africa is one of seven countries that has experienced post-1994 some of the largest increases in arrivals worldwide. It has been rated by the World Tourism Organisation as one of the most promising destinations in Africa (Lewis, 2001). Between 1994 and 1998 the number of international tourists escalated from 700,000 to 1.4m. Since 1999, however, the Mandela tourism boom has waned somewhat and tourism arrivals for 1999-2000 show that growth is levelling off. During 2000, total foreign visitor arrivals fell for the first time since 1986 (Mvoko, 2001). This downturn in overseas arrivals is troubling in light of the special developmental role that South Africa places on tourism as a 'rising star' of the economy.

Nevertheless, it must be recognised that *domestic* rather than international tourism constitutes the heart of South Africa's tourism economy, accounting for 87 per cent of the tourism industry (Page, 1999; DEAT, 2001). Moreover, of the category of 'international' tourist arrivals in South Africa, the largest share of visitors originate from other countries in Africa rather than from Europe or North America, a finding which points to the often overlooked importance of *regional* tourism for South Africa (Ghimire, 2001). A recent World Bank analysis argues that South Africa has the potential both to benefit further as a tourist destination itself and to gain from extension of its tourism industry borders into the rest of southern Africa. With its comparative advantages of natural beauty, temperate climate and diverse attractions combined with its superior infrastructure and transport links, 'South Africa can use such an "integrated" package to differentiate southern African tourism from the rest of the continent, and provide a boost to its own and the regional economy' (Lewis, 2001, p. 85). Overall, the World Bank study on prospects for the economy concludes that 'the case for focusing on tourism as a potential source of growth and employment in today's South Africa is compelling' (ibid., p. 86).

It is against this backcloth that the chapter aims to examine certain major policy debates concerning the development of South Africa's tourism economy and to highlight issues surrounding the country's changing economic geography. The leading themes in the shifting national policy environment for tourism are discussed. Following this, three critical issues are discussed concerning tourism's newfound role as economic driver. In terms of the making of a new economic geography of tourism these issues focus on: (1) pro-poor tourism and the role of tourism in poverty alleviation; (2) the promotion of new tourism growth and investment; and (3) restructuring the tourism space economy.

Changes in Policy Frameworks since 1994

It is often stated that 'tourism development in South Africa has largely been a missed opportunity' (South Africa, 1996, p. 4). Nevertheless, for tourism to develop as South Africa's 'new gold' (The Cluster Consortium, 1999b, p. 3) it is essential that a number of constraints need to be addressed. Since 1994 new policy frameworks have been put in place to support the development of tourism. Most important is the appearance in 1996 of a *White Paper on The Development and Promotion of Tourism in South Africa* (South Africa, 1996) and in 1998 of the *Tourism in GEAR* strategy document (South Africa, 1998). The *White Paper* and the *Tourism in GEAR* strategy link together to provide the key policy foundations for developing the tourism industry in South Africa. A number of subsequent policy documents further elaborate critical dimensions of the new tourism policy framework for South Africa (South Africa, 1999; DEAT, 2000a, 2000b).

The *White Paper on the Development and Promotion of Tourism* is the core of South Africa's new tourism policy. It identifies tourism as a priority for national economic development and a major stimulus for achieving the objectives of the government's GEAR strategy. Several key constraints are seen as limiting the effectiveness of the tourism industry to play a more meaningful role in the national economy. The major constraints relate, *inter alia*, to the fact that tourism has been inadequately resourced and funded by government; the short-sightedness of the private sector towards the nature of the South African tourism product; the limited integration of local communities and previously neglected groups into tourism; inadequate or non-existent tourism training, education and awareness; inadequate protection of the environment through environmental management; poor level of service standards within the industry; the lack of infrastructure in rural areas; the lack of appropriate institutional structures; and the immediate problem of violence, crime and the security of tourists (South Africa, 1996).

The *White Paper* aims to chart a path towards a 'new tourism' in South Africa and offers proposals to unblock these constraints within the context of objectives for reconstruction. In terms of planning, 'the concept of "Responsible Tourism" emerges as the most appropriate concept for the development of tourism in South Africa' (South Africa, 1996, p. 19). Among several features of responsible tourism, four are of special importance. First, that it implies a 'proactive approach by tourism industry partners to develop, market and manage the tourism industry in a responsible manner, so as to create a competitive advantage' (ibid., 1996, p. 19). Second, responsible tourism means responsibility of government and business to involve the local communities that are in close proximity to tourism plant and attractions through the development of economic linkages (e.g. the supply of agricultural produce to the lodges, out-sourcing of laundry, and so on). Third, it implies a responsibility 'to respect, invest in and develop local cultures and protect them from over-commercialisation and over-exploitation'. Finally, it signals a responsibility for local communities to become involved in the tourism industry and to practise sustainable tourism (ibid.).

Six key guiding principles are put forward towards developing responsible tourism in post-apartheid South Africa:

- tourism will be private sector driven;
- government will provide the enabling framework for the industry to flourish;
- effective community involvement will form the basis of tourism growth;
- tourism development will be underpinned by sustainable environmental practices;
- tourism development is dependent on the establishment of co-operation and close partnerships among key stakeholders;
- tourism will be used as a development tool for the empowerment of previously neglected communities and should particularly focus on the empowerment of women in such communities (South Africa, 1996, p. 23).

Building upon the foundations provided by the *White Paper*, the Department of Environmental Affairs and Tourism document, *Tourism in GEAR* aims to create a framework for implementing these policies, particularly within the context of the GEAR macro-economic strategy (South Africa, 1998). This document is of particular interest as in initial post-apartheid economic planning, tourism was something of an afterthought and was omitted entirely from the GEAR framework (Page, 1999). The discovery of tourism's potential as an economic driver is based on several features, *inter alia*, the comparative advantages of South Africa's natural and cultural resources; the fact that South Africa's tourism attractions complement global trends towards alternative tourism; the ability of tourism to attract substantial private sector investment as well as to accommodate small, medium and micro enterprise (SMME) development; the employment-intensive nature of tourism; its potential catalytic role for major infrastructural investment; its ability to stimulate linkages with other production sectors (jewellery, curios); and its value as an export earner (South Africa, 1998). Overall, the vision is 'to develop the tourism sector as a national priority in a sustainable and acceptable manner so that it will significantly contribute to the improvement of the quality of life of every South African' (ibid., p. 4).

Taken together the *White Paper* on Tourism and the *Tourism in GEAR* document signal the need for a collaborative approach within which 'tourism should be led by government and driven by the private sector, and be community-based and labour-conscious' (South Africa, 1999, p. 1).

Among the most critical roles for national government, as a policy and strategic leader for the tourism industry, is that of seeking 'to rectify historical industry imbalances, resulting from a discriminatory political system by promoting tourism entrepreneurship, human resources development, equity and ownership among disadvantaged individuals and communities' (ibid.). Transformation is a critical problem facing the tourism economy and requires expanding the involvement of South Africa's historically disadvantaged black populations (Pillay, 2000). The South African tourism economy is lilywhite in terms of the structure of ownership of leading enterprises (Page, 1999; The Cluster Consortium, 1999b). Indeed, The Cluster Consortium (1999a, p. III-10) reports that white entrepreneurs own roughly 95 per cent of the industry. Thus, an urgent challenge confronting the South African tourism industry is that of 'changing the nature of the South African tourism industry from one that is predominantly white-owned to one that is increasingly owned equitably by the majority of South Africans' (DEAT, 2000b, p. 1). Unless this issue is adequately addressed, there is a danger that tourism's growth will reinforce the concentration of wealth in the hands of whites at the expense of the majority of previously disadvantaged individuals.

Other strategic challenges facing the modern South African tourism industry were indicated in an April 2000 policy document issued by DEAT (2000a). In terms of unlocking South Africa's tourism potential over the next decade a series of strategic challenges are identified that demand co-operation from government, the private sector and communities (ibid., p. 6). First, the imperative to sustain growth in tourism arrivals and in particular in visitor numbers from high-yield tourism sources. Second, the importance of stimulating and supporting emerging tourism entrepreneurs and of maximising opportunities for the SMME sector. Third, to integrate tourism development with strategic frameworks for infrastructure investment. Fourth, to ensure a quality tourism experience and quality products and services. Fifth, to create an overall environment which is conducive to the growth of the tourism industry (DEAT, 2000a).

Finally, The Cluster Consortium (1999b) study stresses a number of fundamental structural challenges that also confront the South African tourist industry. It argues that 'South African tourism development faces a leadership challenge' and notwithstanding the development of elaborate policy frameworks, 'the government has yet to develop a strategy for tourism development that is consistent with South Africa's strong potential in the international and domestic marketplace, and with its imperative for job creation' (ibid., p. 2).

This changed policy environment since 1994 has laid foundations for emerging new economic geographies of tourism in South (and southern)

Africa. Three core issues relating to tourism and economic development are analysed here.

Tourism and Poverty

In the international context, one of the most distinguishing facets of South Africa is the strong commitment made towards tourism assuming a developmental role (South Africa, 1996). More especially, a strong emphasis in South African planning is upon job creation and enterprise development in support of the country's previously neglected black communities. For example, as an integral part of the national government's programmes for poverty alleviation, the DEAT operates a Poverty Relief Programme. The central aim is to catalyse long-term sustainable work opportunities through improving tourism potential, creating new facilities or infrastructure, and encouraging communities to provide better services. DEAT's poverty relief priorities seek to integrate with and firmly support broader national government initiatives. Other important national initiatives that link tourism and poverty alleviation have been launched as part of South Africa's Spatial Development Initiatives (SDI) Programme (Rogerson 2001a, 2001b).

At least four different sets of interventions may be recognised as designed to make tourism a vehicle for poverty alleviation in South Africa. First, substantial investments have been made in improving the infrastructure for tourism development in areas of South Africa with untapped potential. In particular, for many poor rural communities marked by underdevelopment and poverty, access to tourism opportunities is fundamentally constrained by absence of the infrastructure needed to attract investment by tourism developers or to access tourism markets (Viljoen and Naicker, 2000). The initial incorporation of tourism in the SDI programme, which aims at unblocking investment opportunities, was associated with commitments made to poverty alleviation (Rogerson, 2001b). The Lubombo SDI and the Wild Coast SDI represent two good examples of infrastructure improvement which is linked to the opening up of new tourism opportunities for rural communities in these areas. In addition, the Maputo Development Corridor, the West Coast Investment Initiative and even the Gauteng SDI have also incorporated substantial commitments towards infrastructural improvements that allow poor communities to participate in tourism (see Rogerson, 2001a, 2001b; Rogerson and Sithole, 2001).

A second stream of interventions that link tourism and poverty alleviation relates to the search for new market niches and the development of new tourism products that involve poor communities. In relation to tourism product development, the attention given to cultural tourism offers considerable job and enterprise opportunities for many poor communities. For example, cultural villages offer both direct and indirect employment and entrepreneurship opportunities in many rural areas of South Africa (Jansen van Veuren, 2001). The most important indirect spin-offs relate to the selling of rural handicrafts as well as cultural performances (Rogerson and Sithole, 2001). Another aspect of cultural tourism is the initiatives to promote 'township tourism' in places such as Soweto, Alexandra (Johannesburg) or Khayelitsha (Cape Town). In the growth of these areas as new tourism spaces the focus is often on the political struggle for democracy in South Africa. Once again, this new alternative form of tourism creates a range of indirect livelihood opportunities, including the provision of food, drinks as well as craft retailing.

Nature tourism is a further critical area for expansion that is linked to poverty alleviation in many rural areas (Viljoen and Naicker, 2000). Certain South African ecotourism enterprises, such as Wilderness Safaris and Conservation Corporation Africa, have demonstrated a commitment to expand the benefits of tourism development to local communities. Promotion of small businesses and the informal sector as well as community involvement in the ecolodges is an important element of the activities of these enterprises. More broadly, in the investment projects promoted through the SDI programme there is a strong 'empowerment' component in tourism development with the focus upon the maximisation of the benefits of private sector investment for local communities (Rogerson, 2001c). The key factor in the success of these pro-poor initiatives is, however, the financial viability of the core business (Poultney and Spenceley, 2001).

Third, the improvement of the skills base of poor communities through various training initiatives is another important policy intervention. DEAT has facilitated the establishment of the Tourism and Hospitality Education and Training Authority (THETA) which functions as the overarching training authority for the tourism industry. The THETA is devising a National Training Strategy for accreditation, setting of standards and disbursement of levies. The critical issue of job creation is addressed through a Learnerships programme which DEAT facilitates. Most of these learnerships target students and unemployed youth and aim to improve their employment prospects in tourism through a combination of structured learning and structured workplace experience (DEAT, 2000b). Skills training focuses upon a range of occupational categories in which an expansion of job opportunities is anticipated relating to tourism expansion in both rural and urban areas of South Africa.

The introduction of various new forms of ownership and institutional arrangements for tourism projects is a final vital aspect for poverty alleviation. The core emphasis has been upon direct involvement of communities in the ownership and operations of tourism projects. The South African experience shows that if communities have land ownership as an asset base there is a potential for strong economic and social empowerment with considerable benefits flowing to poor communities. Correspondingly, where the land rights situation is unclear, communities may still benefit from tourism, albeit to a lesser extent. Through the SDI policy programme much attention was given to the development of new 'empo-tourism' models which sought to move away from a corporate responsibility approach to one that is anchored on the underlying rights and abilities to add value to such deals (Elliffe et al., 1998). These new models are being implemented through so-termed 'Community Public Private Partnerships' which aim to revitalise depressed rural economies through linking 'resource rich' communities with private investors interested in the sustainable utilisation of natural assets (Elliffe, 1999). One of the most advanced CPPP projects is the Makuleke tourism initiative in Northern Province which arose as a consequence of the victory of the local community in a land restitution case. The Makuleke community is involved in a joint venture with the private sector in developing new game lodges and other tourism products in an area which is functionally part of Kruger National Park (Mahony and van Zyl, 2001).

Taken together these several ongoing interventions surrounding tourism development and poverty alleviation suggest that, over the next decade, South Africa may emerge as one of the international leaders in, what is described as, 'pro-poor tourism' (Ashley et al., 2000). The essential focus of pro-poor tourism strategies is to unlock opportunities for economic gain, livelihood benefits and decision-making for the poor (Ashley et al., 2001).

Tourism Growth and Investment

The promotion of new tourism growth and investment in tourism plant and product development is of central importance, not least for achieving certain goals of poverty alleviation. The restructuring of SATOUR, the national tourism organisation, accompanied the formulation and implementation of certain positive international marketing strategies and promotional campaigns. The personnel and management composition of SATOUR has been altered and the organisation is set to market South Africa as a whole 'rather than only the traditional features that have been mainly rich or white icons' (DEAT, 2000b, p. 5). None the less, The Cluster Consortium study (1999a, 1999b) stresses the need for South Africa to break from past marketing approaches which were seen as fragmented, unco-ordinated and short-term. Instead, it recommends that South Africa establish a co-ordinated and integrated strategic destination-marketing framework which adheres to global best practice approaches in destination marketing. Essential elements of such an approach are development of a clear and distinct branding for South Africa in international tourism and ensuring a targeted approach towards international marketing which is based on sound research, market analysis and segmentation.

Traditionally, South African tourism marketing has focussed on the 'Big Six' markets of the United Kingdom, Netherlands, Germany, France, Italy and USA, which make up 37 per cent of all overseas arrivals (DEAT, 2001). New marketing initiatives are starting to focus more on selling South Africa in emerging markets, such as India, Japan, China and countries of south-east Asia (DEAT, 2001). The 2001 action plan for tourism centres on two international campaigns to be run during the year in which marketing efforts are geared to 'matching the efforts of competitor countries as well as opening up new markets' (Dobson, 2001). Further marketing of South Africa is undertaken as part of the activities of the Regional Tourism Organisation of Southern Africa. Indeed, it is acknowledged increasingly that the SADC countries are not competitors; rather their tourism strengths should be harnessed to build collectively the tourism economy of southern Africa. Lastly, the importance of marketing for domestic tourism is not overlooked.

The DEAT has been engaged in formulating a tourism investment promotion strategy with the goals particularly of creating long-term jobs in the industry, influencing transformation through increasing investment into projects from disadvantaged individuals and/or communities, and promoting regional co-operation in southern Africa through co-ordinated regional investment promotion programmes. In partnership with the Department of Trade and Industry, the DEAT is seeking to develop a holistic and integrated support package for investment into tourism, consisting of training, market linkages, mentorship assistance for emerging tourism entrepreneurs and access to affordable finance (DEAT, 2001). This investment drive seeks to build upon those initiatives undertaken during 1999-2000 designed to anchor the building of a competitive tourism nation. Of central importance was the SA Welcome Campaign which aimed to increase awareness amongst South Africans of the economic significance of tourism. In addition, attention was devoted to improving service and hospitality through the Welcome Host programme for customer training.

This draws on similar initiatives used in Canada, the UK and Australia teaching people 'how to go the extra mile for tourists' (DEAT, 2001).

Essentially, these changes are part of broader efforts designed to 'professionalise' the tourism industry in South Africa and to extend its markets beyond those of the traditional offering of 'sunny South Africa'. Indeed, beyond the traditional fare of beach holidays, family visits, business tourism and game visits, South Africa is now marketed as 'a world in one country'. New brand 'South Africa' also offers cultural tourism, ecotourism, casinos and theme parks, historical township and battlefield tours, golf tours, World Heritage Sites, white water rafting, backpacking as well as venues for major conferences, events and exhibitions (Addison, 2001; DEAT, 2001).

Since 1994 a substantial expansion has occurred in investment in tourism, notwithstanding both the high risks of the industry and the high costs of capital (Business Map SA, 1999). Nevertheless, 1997 represented the peak year and since 1998 decreased tourism investment is linked to high interest rates and the lack of major investment incentives for tourism. Indeed, it is striking that several major South African tourism enterprises, such as Sun International, Southern Sun and Protea, are involved in larger investments in new tourism plant outside of South Africa, particularly in Zambia and Mozambique. Overall growth in South African tourism investment has been strongly driven by the gaming sector. A restructuring of the casino industry has taken place from the old apartheid pattern of casinos being ghettoised to the former homelands. A total of 40 new licences are to be awarded on a provincial basis and under the new licensing dispensation a spatial shift of casino investments has occurred into the country's major urban centres. Although gaming has been the most significant sector for new investment, the accommodation sector remains the dominant sector for tourism investment. During the post-apartheid period, a number of new foreign investors have established themselves in the South African tourism industry (e.g. Mercure, Formule One, Sheraton) whilst established tourism enterprises (e.g. Protea, Southern Sun) also expanded their operations to take advantage of increases in both the domestic and international tourism markets.

Several factors are suggested as blockages on new private sector investment in the South African tourism economy. First, there is widespread concern about unacceptable levels of safety and security, not only in South Africa but the subcontinent and Africa as a whole. Negative perceptions concerning personal safety represent one of the major threats to South Africa's country's tourism industry (Ferreira, 1999; The Cluster Consortium, 1999a; Ferreira and Harmse, 2000). A second associated factor concerns perceptions regarding health threats and health facilities. In particular, potential investors in the tourism economy have been alarmed by the risks of diseases and viruses such as malaria, HIV and cholera (Lewis, 2001). Third, many foreign investors express worries about issues of 'political instability' (Business Map SA, 1999, p. 7), a situation that has been exacerbated by conflicts and political conditions in Zimbabwe, Angola and the Democratic Republic of Congo (DEAT, 2001). Fourth, investment has been negatively affected by a volatile demand for products, both in terms of the tourism product supplied and by changing global demands (Business Map SA, 1999). Moreover, high-profile events such as terrorist bombings in Cape Town are detrimental for the South African tourism economy as a whole. Fifth, relative to other sectors for investment, the tourism sector historically has not performed well as a sector for investment in South Africa (Business Map SA, 1999).

Finally, investment in tourism also has been affected by the meagre financial incentives offered by national government. Although there have existed certain investment incentives designed for tourism (Investment SA, 1998), a recent comprehensive review suggested that these 'are limited in South Africa' (KPMG, 1999a, p. 17). As compared with the many investment packages offered to potential manufacturing investors those for tourism have been small. Moreover, given the challenges facing tourism in South Africa and its importance for future economic development, the incentives that have been provided are not conducive to the position of the tourism sector. During 2000, national government initiated a major review of existing investment incentives and in association with the Department of Trade and Industry, a set of dedicated tourism incentives were developed in order to stimulate the tourism economy (DEAT, 2000a, p. 13). The new tourism incentives announced in May 2001 target qualifying operators who invest up to R100m. in assets for new projects or expansion of existing tourism plant (Zhuwakinyu, 2001).

Transforming the Tourism Space Economy

With the implementation of this range of new policy initiatives for tourism and poverty alleviation alongside those for promoting new tourism growth, there are changes taking place in the spatial organisation of tourism in South Africa. The clear direction of these spatial shifts is towards seeking to spread more widely the benefits of the growing tourism economy across South Africa.

Geographically, investments in tourism have been concentrated upon opportunities in the Western Cape (particularly the Cape Town area) and Gauteng, which are areas that are perceived as a relatively low risk for investors. Further, these two provinces enjoy the highest levels for disposable incomes across South Africa and thus have been attractive for investors in the casino/gaming sector. Outside of Gauteng and Western Cape, it is notable that in the perceived higher risk investment areas, such as the SDI areas, large-scale tourism investment flows so far have been limited (Rogerson, 2001a). The key investors in the South African tourism economy are domestic rather than foreign investors. For the period 1994-98 it was estimated that domestic investment accounted for 85 per cent of total commitments in the tourism economy. Spatially, investments are focussed in three clusters, namely around casinos and entertainment complexes, the business tourism economy of Gauteng, and leisure tourism investment in the Western Cape. Overall, The Cluster Consortium (1999a, II-16) records that in terms of domestic investors the award of casino licences together with support for business travellers has meant that Gauteng has garnered 'a large chunk of investment'.

The existing tourism space economy of South Africa is thus highly uneven and polarised. For international tourists the typical route is of entry at the gateway of Johannesburg International Airport (JIA), overnight in Johannesburg, onward to Kruger National Park, south to Cape Town, on to the Garden Route and back to JIA. Both the geographical supply and demand for accommodation in South Africa is highly concentrated in three main areas of the country, namely metropolitan Cape Town, Durban and Gauteng (KPMG, 1999b). The Gauteng tourism node is primarily led by business tourism. The Durban and Cape Town tourist clusters are based upon mainly domestic leisure tourism with international leisure tourism of particular and growing importance in Cape Town and its environs. Outside of these urban centres the only other nationally significant nodes for tourism development are in Mpumalanga, the gateway to South Africa's game parks, and along the Cape Garden Route. This geographically polarised pattern of tourism development inevitably means that the benefits of tourism in South Africa are currently distributed in a spatially uneven manner with few benefits or opportunities flowing outside of these major tourism areas.

It is evident that one necessary element in the transformation of the South African tourism industry is the need to change its existing spatial structure. Although South Africa at present does not have a formalised spatial tourism strategy, 'the key elements which are needed in order to create one have been identified, namely existing demand and supply, future potential and areas in need of development' (KPMG, 1999a, p. 15). New investments and tourism flows are sought outside of the traditional heartlands for tourism development in South Africa and the SDI programme has been important in seeking to open up the potential of many areas. Other new initiatives that may spread tourism flows include bold cross-border initiatives for establishing Trans-frontier Parks and Trans-frontier Conservation Areas. Overall, in order to maximise economic and social benefits from tourism, there is a clear requirement that tourists should be dispersed further away from the established tourist destination areas into other parts of South Africa. Many of these new tourism spaces are important focal points for addressing rural poverty.

One critical factor blocking the spatial transformation of the tourism economy is, however, the lack of essential infrastructure which has been an undoubted deterrent to many potential investors. The provision of infrastructure as a means to 'debottleneck' investment opportunities in areas of tourism potential was one element of the SDI programme, especially in the Lubombo and Wild Coast SDIs in which tourism is the lead sector. In important DEAT policy documents, nineteen so-termed 'priority areas for tourism infrastructure investment' or 'PATIIs' are defined across South Africa (KPMG, 1999a). The nineteen identified PATIIs represent geographical areas which exhibit high potential for tourism development and are planned to form the base for tourism product development in South Africa (DEAT, 2000b, p. 7). Nevertheless, these areas presently are acknowledged to have a weak infrastructural base (KPMG, 1999a). Essentially, the PATIIs provide a spatial matrix for the channelling of new investment into the South African tourism economy in order to maximise its impact for developing tourism in areas of greatest untapped potential. Although the majority of these PATIIs are situated in rural areas, a number of them are located in South Africa's major metropolitan areas, including Cape Town, Durban and Johannesburg (ibid.). For example, Soweto is identified as one of the PATIIs in urban South Africa with infrastructural improvements seen as one element in unlocking the area's tourism potential. Overall, the network of PATIIs function as important elements for transforming the future tourism space economy and achieving a greater geographical dispersal of tourists and tourism business opportunities across South Africa.

Another key element in transforming the spatial structure of the South African tourism economy is maximisation of the impacts of *local*-level initiatives for tourism-led economic development. Over the last decade, a number of tourism-led initiatives have been launched in South Africa as a basis for promoting local economic development (Rogerson, 2002a, 2002b). Until the 1994 democratic transition, few opportunities existed for local initiatives for tourism-led local economic development (LED). With the growth of the tourism economy as a whole, many South African localities are re-focussing their developmental initiatives away from their traditional sectoral focus on manufacturing and instead seeking to promote tourism as a lead sector (Rogerson, 2000b, 2002a). Of particular note have been a set of initiatives for developing 'route tourism' in which a number of local authorities seek to co-operate together in order to compete more effectively on a collective basis in tourism markets. Among the most well-known of these route tourism initiatives in South Africa are the Midlands Meander in KwaZulu-Natal province and the Highlands Meander in Mpumalanga.

Several critical issues confront the successful development of these unfolding local initiatives for tourism-led LED. The most important issue is, perhaps, that of ensuring that benefits from tourism spread into poor communities that did not benefit during the apartheid years (Mathfield, 2000). Patterns of employment creation, business linkage and subcontracting of work to small enterprises are the issues of central importance in achieving a wider spread of benefits than occurred in the past (Rogerson, 2000a, 2002a, 2002b).

Acknowledgements

Thanks are due to the National Research Foundation, Pretoria for research support.

References

Addison, G. (2001), 'Growing Pains', *Siyaya*, vol. 7 (Summer), pp. 6-12.

Ashley, C., Boyd, C. and Goodwin, H. (2000), *Pro-poor Tourism: Putting Poverty at the Heart of the Tourism Agenda*, Natural Resources Perspectives No. 61, Overseas Development Institute, London.

Ashley, C., Roe, D. and Goodwin, H. (2001), *Pro-Poor Tourism Strategies: Making Tourism Work for the Poor*, Pro-Poor Tourism Report No. 1, Overseas Development Institute, London.

Business Map SA (1999), 'Investment in Tourism', unpublished report prepared for the Cluster Consortium, Johannesburg.

DEAT (2000a), *Unblocking Delivery on Tourism Strategy by Government Departments*, Department of Environmental Affairs and Tourism, Pretoria.

DEAT (2000b), 'Transforming the South African Tourism Industry', unpublished report, Department of Environmental Affairs and Tourism, Pretoria.

DEAT (2001), *Annual Review 2000-01: Tourism – The Way to Go*, available at www.environment.gov.za.

Dobson, W. (2001), 'Driving Growth – from Macroeconomic Stabilisation to Microeconomic Reform', *Sisebenza Sonke*, Issue 1, Article 3.

Elliffe, S. (1999), 'Guidelines for the Release/Development of Dormant State or Community Assets for Eco-tourism Development in the Context of Community Involvement, Land Issues and Environmental Requirements', unpublished paper presented at the community Public Private Partnerships Conference, Johannesburg, 16-18 November.

Elliffe, S., Rutsch, P. and de Beer, G. (1998), *Allocating Community Benefits: Institutional Options in Support of Empo-Tourism Models*, Working Paper No. 14, Development Policy Research Unit, University of Cape Town.

Ferreira, S.L.A. (1999), 'Crime – a Threat to Tourism in South Africa', *Tourism Geographies*, vol. 1, pp. 316-24.

Ferreira, S.L.A. and Harmse, A.C. (2000), 'Crime and Tourism in South Africa: International Tourists Perception and Risk', *South African Geographical Journal*, vol. 82, pp. 80-85.

Ghimire, K.B. (2001), 'Regional Tourism and South-South Economic Co-operation', *Geographical Journal*, vol. 167, pp. 99-110.

Investment SA (1998), *Investing in Tourism*, Investment SA and Department of Environmental Affairs and Tourism, Sandton.

Jansen van Veuren, E. (2001), 'Transforming Cultural Villages in the Spatial Development Initiatives of South Africa', *South African Geographical Journal*, vol. 83, pp. 137-48.

KPMG (1999a), 'Review of Tourism Investment Incentives: Perspectives and Issues', unpublished report for the Department of Trade and Industry, Pretoria.

KPMG (1999b), 'Review of Infrastructure in Support of International and Domestic Tourism Development: Final Report - Methodology, Identification and Mapping', unpublished report for the Department of Environmental Affairs and Tourism, Pretoria.

Lewis, J.D. (2001), *Policies to Promote Growth and Employment in South Africa*, Discussion Paper no. 16, Informal Discussion Papers on Aspects of the Economy of South Africa, The World Bank, Washington DC.

Mahony, K. and Van Zyl, J. (2001), *Practical Strategies for Pro-Poor Tourism: Case Studies of Makuleke and Manyeleti Tourism Initiatives*, Pro-Poor Tourism Working Paper No. 2, Overseas Development Institute, London.

Mathfield, D. (2000), 'Impacts of Accommodation and Craft-based Tourism on Local Economic Development: the case of the Midlands Meander', unpublished Masters dissertation, University of Natal, Durban.

Mvoko, V. (2001), 'Report Shows all is not Well in SA's Tourism Sector', *Business Day*, 5 September.

Page, S. (1999), 'Tourism and Development: the Evidence from Mauritius, South Africa and Zimbabwe', unpublished report prepared for the Overseas Development Institute, London.

Pillay, S. (2000), 'Towards a Strategy to Transform Tourism in South Africa', unpublished report, Department of Environmental Affairs and Tourism, Pretoria.

Poultney, C. and Spenceley, A. (2001), *Practical Strategies for Pro-Poor Tourism: Wilderness Safaris South Africa: Rocktail Bay and Ndumu Lodge*, Pro-Poor Tourism Working Paper No. 1, Overseas Development Institute, London.

Rogerson, C.M. (2000a), 'Subcontracting in the Tourism Industry: International Best Practice and the Nature of Subcontracting between Large Hotel Establishments and Small Businesses in South Africa', unpublished report prepared for The Development Bank of Southern Africa and Ntsika Enterprise Promotion Agency, Midrand and Pretoria.

Rogerson, C.M. (2000b), 'Local Economic Development in an Era of Globalisation: the Case of South African Cities', *Tijdschrift voor Economische en Sociale Geografie*, vol. 91, pp. 397-411.

Rogerson, C.M. (2001a), 'Tourism and Spatial Development Initiatives: the Case of the Maputo Development Corridor', *South African Geographical Journal*, vol. 83, pp. 124-36.

Rogerson, C.M. (2001b), 'Tracking South Africa's Spatial Development Initiatives', in M. Khosa (ed), *Empowerment Through Economic Transformation*, African Millennium Press, Durban, pp. 249-69.

Rogerson, C.M. (2001c), 'Investment-led Entrepreneurship and Small Enterprise Development in Tourism: Lessons for SDIs from the International Experience', *South African Geographical Journal*, vol. 83, pp. 105-14.

Rogerson, C.M (2002a), 'Tourism and Local Economic Development: the Case of the Highlands Meander', *Development Southern Africa*, vol. 19, in press.

Rogerson, C.M. (2002b), 'Tourism and Local Economic Development: the South African Experience', *Urban Forum,* vol. 13, in press.

Rogerson, C.M. and Sithole, P.M. (2001), 'Rural Handicraft Production in Mpumalanga, South Africa: Organization, Problems and Support Needs', *South African Geographical Journal*, vol. 83, pp. 149-58.

South Africa (1996), *White Paper on the Development and Promotion of Tourism in South Africa*, Department of Environmental Affairs and Tourism, Pretoria.

South Africa (1998), *Tourism in Gear: Tourism Development Strategy 1998-2000*, Department of Environmental Affairs and Tourism, Pretoria.

South Africa (1999), *Institutional Guidelines for Public Sector Tourism Development and Promotion in South Africa*, Department of Environmental Affairs and Tourism, Pretoria.

The Cluster Consortium (1999a), 'South Africa's Tourism Challenge: A Profile of the Tourism Cluster', unpublished report prepared for the Tourism Clustering Initiative.

The Cluster Consortium (1999b), 'The South African Tourism Cluster: Strategy in Action', unpublished report prepared for the Tourism Clustering Initiative.

Viljoen, J.H. and Naicker, K. (2000), 'Nature-based Tourism on Communal Land: the Mavhulani experience', *Development Southern Africa*, vol.17, pp. 135-48.

World Travel and Tourism Council (1998), *South Africa's Travel and Tourism: Economic Driver for the 21st Century*, World Travel and Tourism Council, London.

Zhuwakinyu, M. (2001), 'Tourism Incentive Applications Flow', *Engineering News*, 10 August.

6 South African Finance across Time and Space

PATRICK BOND

Introduction

South Africa has a well developed, mature financial sector dominated by fewer than a dozen major institutions (banks, insurance companies and other financial firms). In surveys by the World Economic Forum, South Africa regularly rates amongst the most competitive of middle-income countries' financial systems. The typical phrase used to describe South African finance is 'sophisticated', as a result of the institutions' capital intensity, marketing prowess and range of high-end products for corporate clients and wealthy consumers. Geographically, there can be found branches of virtually all the major financial firms in most urban centres, including small towns with populations as low as a few thousand people.

Yet by all accounts, the financial markets have not fostered an effective, efficient mobilisation of savings and investment resources to benefit either the vast majority of the country's population or the vast majority of small (often informal) enterprises. There remains an apartheid residue in the financial system that corresponds to deep-seated historical processes of race, gender and class oppression. The location of branches is invariably biased towards traditionally white areas of town, and desegregation of inner cities is too often accompanied by 'redlining' (geographical credit discrimination) and the closure of financial institutions. While compared with the pre-1994 era, race and gender are less likely to be direct determinants of whether potential customers are excluded, class is an especially powerful barrier to financial-system access in post-apartheid South Africa, given that even opening a savings account is a privilege reserved for the formally employed. Moreover, the country's financial institutions are not immune to market failure, and many small banks and financial firms have gone bankrupt over the past fifteen years.

In short, finance remains one of the key contributing forces to South Africa's *uneven development*, which in turn is amongst the world's most extreme and painful. The contribution of finance to inequality and underdevelopment is a general problem across the world. But South African financial geography also deserves a particularly critical analysis across time, given that formal institutions have operated since the late eighteenth century. The past two decades or so witnessed an amplification of both power and vulnerability in the financial sector, with notable implications for spatial and social development during political transition. Unfortunately, in spite of a strong mandate in the African National Congress (ANC) 1994 campaign platform (the *Reconstruction and Development Programme*) to regulate domestic finance, the primary domestic post-apartheid financial processes have exacerbated problems associated with unequal and inefficient economic geography. Part of the problem stems directly from international economic processes, which since 1995 have battered South Africa with extremely uneven inward and outward flows of portfolio finance. Growing international public policy concern over the economic geography of speculative finance is reflected in periodic demands for reform, including from South African grassroots, labour and church activists.

In short, this chapter considers how the financial system contributed to apartheid geographical forms, to South Africa's development trajectory, to the demise of apartheid and to post-apartheid economic processes. The centrality of finance across space and time will be obvious enough. But this is no more permanent a barrier to a geography of social justice than the power relations that sustained apartheid – even if during the 1980s-90s era of free-market 'neoliberalism', the worsening of uneven financial development appeared as a formidable barrier to authorities and citizens alike.

Finance and Unevenness in Economic Geography

Financial markets and uneven geographical development are inextricably linked (Bond, 1999a, 1999b). As just one reflection, by the late 1990s, the leading practitioner of international currency speculation, George Soros (1997, p. 4), confirmed that cross-border financial flows were out of control:

> The private sector is ill-suited to allocate international credit. It provides either too little or too much. It does not have the information with which to form a balanced judgment. Moreover, it is not concerned with maintaining macroeconomic balance in the borrowing countries. Its goals are to maximise profit and minimise risk. This makes it move in a herd-like fashion in both directions. The excess always begins with over-expansion, and the correction is always associated with pain.

Is finance necessarily a disruptive force in economic relations because of its herd-like tendencies? At a theoretical level, David Harvey (1989, p. 176) has forcefully argued that the relationship between finance and uneven development depends upon the prevailing balance of forces: 'Money creates an enormous capacity to concentrate social power in space, for unlike other use values it can be accumulated at a particular place without restraint'.

Finance is, therefore, much more than a mere service enterprise with specific locational features, to be incorporated into ordinary models of the space-economy. In traditional geography, the location of financial institutions and the distribution of their physical lending and deposit-taking functions had long been considered effective barometers of economic vitality or of other productive sector activities less easily traced. The obvious failure of such an approach, though, is that it assumes, rather than questions, the underlying rationality of financial geography. In the context of speculative financial markets and the rise of financial power, which shaped both South Africa's history and contemporary options, this assumption is immediately suspect.

What generates the power of finance to amplify uneven development? Briefly, over long-term economic cycles, the logic of capital accumulation entails periods of excess capital investment, and hence lower rates of profitability in the economy's productive sectors. There logically follows a shift of investible resources out of plant and equipment and into higher-earning financial assets. The denouement – financial crisis – typically also marks the end of a long-term cycle of accumulation (e.g., mid-1810s, late 1840s, early 1870s, mid-1890s, late 1910s, early 1930s).

The depth of historical evidence for such phenomena varies depending on the contingencies of different national financial institutions and systems. But across the world during the last quarter of the twentieth century, uneven financial development became internationalised. Financial markets offered extremely high returns on speculation (in shares, real estate, currencies, exotic financial instruments, etc.), and from high interest rates paid on financial assets. To illustrate, profit rates plus salaries in the US Financial, Insurance and Real Estate sector (as a percentage of gross investment) soared from 20-30 per cent returns during the 1950s-70s up to the 35-45 per cent range during the 1980s-90s, and the 'rentier' share (i.e., interest plus dividends) of the US corporate surplus (i.e., pretax profits plus interest) rose from levels of 20-30 per cent during the 1950s to 30-40 per cent during the 1960s-70s, to 50-70 per cent during the 1980s-90s (Henwood, 1998, p. 59, pp. 73-77). Reflecting the speculative movement of money, by the late 1990s daily currency market activity across the world amounted to $1.5 trillion, only a tiny fraction of which was used for trade-

or investment-related transactions. The entire volume of global trade in 1998 - \$6.5 trillion - could have been financed through just 4.3 days worth of forex market turnover (Bank for International Settlements, 1999).

Across a variety of scales, uneven development is generally accentuated during periods such as the late twentieth century when financial institutions increase their range of movement, the velocity and intensity of their operations, and simultaneously, their power over debtors (whether companies, consumers or governments). But specifying how this happens, within a multi-faceted relationship between financial power and uneven spatial development operating at different scales, is challenging, as witnessed by many telling examples from South African economic history.

A History of Uneven Financial Development

The history of financial domination of South Africa's geographical development begins with the onset of substantial international financial flows in the early nineteenth century, followed by the founding of the Standard Chartered Bank in the mid-nineteenth century, and the rise of diamond and gold as sites of mining-financial power, speculation and international investor interest (see, for example, Bond, 1998; Henry, 1963; Mabin, 1984, 1985, 1986, 1989; Schumann, 1938). The most important subsequent periods for the consolidation of the South African financial sector were:

- the massive centralisation of financial-mining capital during the 1880s, and the political implications that lasted through the South African Anglo-Boer War of 1899-1902;
- the reassertion of local control during the 1910s and financial restructuring of local economic geography through the 1920s;
- the international financial collapse during the 1930s and gold-based recovery of the 1930s-40s;
- the rise of Afrikaner finance during the 1930s-50s and the financing of post-war development;
- the contemporary rise of finance.

Although diamonds (1867) and gold (1886) were to offer a site for settler-capital's own accumulation on a vast scale, a broader context of financial power and vulnerability had already been established in the international markets. Ian Phimister (1992, p. 7) contends that by 1885, the political realignment of the African continent emanated from 'capitalism's uneven development during the last third of the nineteenth century, particularly the City of London's crucial role in mediating the development of a world economic system'. As a result, in South Africa the growing mining houses and 'imperial' banks (those with strong British ties) controlled most sites of substantial capital accumulation. A variety of smaller district banks reliant upon small white-dominated capitalist agriculture foundered badly, while the only modes of saving that were available to the majority of people were informal, such as the *stokvels* which emerged in Transkei in the mid-nineteenth century as women sought to establish mutual-aid coping mechanisms to survive the migrant labour system.

With large-scale accumulation came concentration and centralisation of capital. From 1910 to 1926, South Africa's banking system was reduced from seven banks to just two big London banks – Standard Chartered and Barclays – and the smaller Netherlands Bank of South Africa, then headquartered in Amsterdam. But simultaneously, the erosion of the international gold standard – the system which rendered local and British currency directly convertible to gold – weakened the power that the City of London exerted over South Africa's financial system (Ally, 1994). London banks came under extreme war-time and post-war stress due to inflation, the devaluation of the British pound and the rise of New York City as a competitor. J.P. Morgan's New York-based financial empire gained a toehold in South Africa through its role in the founding of Ernest Oppenheimer's Anglo American Corporation. Ultimately, South African authorities decided, the only solution to the financial uncertainty was to create a local Reserve Bank to act as a guarantor for the banks and the South African currency. In the ensuing struggle over the character of banking regulation, the Reserve Bank was essentially put under the direct ownership and control of bankers, unlike other countries where the state owned the central bank (Gelb, 1989).

The weakening of London's links to South Africa opened space for local elites to influence financial and monetary policy. The Reserve Bank's first big challenge was a bail-out of the National Bank, a victim of the financial chaos of the early 1920s. The rescue was facilitated by the Bank of England and by the conclusive rescue of the National Bank in a 1926 takeover by the Anglo-Egyptian Bank and the Colonial Bank, the result of which was the formation of Barclays Bank (Barclays, 1938).

By the fateful year 1929, local bankers were extremely bullish, witnessed by the rise in their ratio of loans to deposits from 63 per cent in 1926 to 85 per cent in 1930. As often happens just before a fall, overproduction of agricultural goods became rife, and the government intervened with increasingly protectionist policies. Imports of wheat, flour and sugar were discouraged, and a Marketing Board was established to support South African exports. And when Standard Bank came under pressure to export funds to its London office, the Reserve Bank imposed a levy for bank remittances and raised the interest rate it charged local banks. In addition, state intervention was required on behalf of rural whites, especially laws supporting debtors' rights. White workers and displaced farmers made a series of proposals for rural credit co-operatives and for municipal banks in Johannesburg and Durban during the mid-1930s.

At the outset of the Great Depression, the value of the South African currency remained high relative to other currencies because of the gold standard, and exports suffered. At the same time, investors were shifting enormous amounts of money out of South Africa (£20m. in 1932). By the end of 1932, the tensions were overwhelming and the country's social fabric was tearing, so mining houses led the charge to abandon the gold standard and devalue the currency. The first result was that vestiges of speculative financing again appeared from abroad, pushed away from Europe and the United States by the deepest depression in capitalist history. Nevertheless, over the subsequent fifteen years, the South African economy was relatively isolated from international manufacturing trade, and thus financing was increasingly directed towards the nascent local manufacturing industry.

In a manner Andre Gunder Frank observed occurring elsewhere on the global economic periphery and which helped generate many of the insights of the 'dependency school', manufacturing grew in inverse relation to the strength of trade and financial flows in the international economy (Frank, 1967). Thanks in part to its trade/financial delinking, South Africa spent the period from 1933 to the 1940s growing faster (8 per cent average GDP increase per annum), more evenly across sectors, and with larger relative wage increases for blacks (from 11 per cent to 17 per cent of the total wage bill), than at any other time in the twentieth century (Nattrass, 1981, Appendix). Later, as South Africa reintegrated into the world economy, racial biases were amplified, for example the black wage share stagnated, reaching just 21 per cent by 1970.

But age-old conflicts between English-speaking financial elites and Afrikaner farmers remained even during prosperity (Gilliomee, 1989; Keegan, 1986; O'Meara, 1983). The legacy of Afrikaner nation-building was partially a function of funds mobilisation on ethnic lines, highlighted by the role of the Broederbond in setting up banks during the late nineteenth century and the 'Economic Movement' of the 1930s and 1940s, which allowed Sanlam, Volkskas, Federale Volksbeweging, Saambou and other ethnic institutions to achieve a sufficient branch network, asset base and financial sophistication to benefit from state patronage following the 1948 National Party electoral victory.

During the 1940s, English-speaking financiers continued to resist the farmers' state subsidies and took advantage of their overproduction and land speculation problems. In 1944, at the Bretton Woods conference – attended by a South African delegate who sided with the US against Britain's John Maynard Keynes and other representatives of debtor nations – an agreement regulating international finance through a semi-gold standard was announced. The mining houses and English-speaking businesses benefited enormously, and the Johannesburg stock market boom of 1946 doubled the number of companies listed. Access to international capital, organised by the local mining houses and stock market immediately after the war, was checked only briefly by hollow Afrikaner nationalist threats of nationalisation in 1948. The London banks helped spur a variety of new financial innovations, including accommodation of corporate investment needs by emerging money markets and of housing needs via building society expansion. A quarter of the mining industry funds were raised from mining Trust Funds in the US and Switzerland, while 7 per cent came from new local financing sources such as the government's National Finance Corporation, founded in 1949 to gather and deploy corporate savings. The rest was sourced from British financial institutions, mainly banks, insurance companies, pension funds, investment and trust units, and various other institutional investors.

The post-war financing arrangements reflected the broader process of concentration and monopoly control, and according to Duncan Innes (1984, p. 150), 'was the clearest form yet of the merging together of bank capital and productive capital – that is, of the emergence of the phase of finance capital. It was thus during the late 1950s that South African capitalism *as a whole* – and not just specific sectors of the economy – displayed the first clear signs of having entered the monopoly phase of its evolution'.

Regulation of the national financial structure would also need to adapt to keep pace with developments. A 1964 banking law allowed banks and building societies greater depth and reach. Funds available to the banking sector soared, and financing on the Johannesburg Stock Exchange was boosted dramatically, until a crash in 1969. Then, fuelled by the dramatic rise in the international price of gold once the US ended its Bretton Woods-era linkage to the dollar in 1971, an inordinate amount of capital was subsequently attracted into geographical expansion over the subsequent decade. Vehicles included the internationalisation of the mining finance houses and the enormous boom in construction. Major state infrastructure projects also involved a great deal of foreign borrowing: nearly a quarter of government parastatal investment from 1972-78 was funded through international capital markets (Bond, 1991).

The forging of such financial links drew the South African economy more tightly into the global economy. In turn, this led to such financial dependence that by the 1970s the international anti-apartheid movement discovered that it represented the country's Achilles Heel, and hence began to focus sanctions pressure on international banks.

Financial Power/Vulnerability during Political Transition

After the Soweto uprising in 1976, Pretoria gained access to International Monetary Fund loans amounting to nearly $2bn. (until borrowing rights were cut in 1983, due to anti-apartheid protest). However, when the 1979-81 gold boom had run its course, foreign banks finally lost confidence in apartheid and in 1985 agreed to cut credit lines for all but short-term trade finance. Unable to roll over the vast loans contracted by private sector borrowers, especially the large US banks that came under the greatest activist pressure, Pretoria was ultimately forced to call a 'debt standstill' and refused to make repayments on more than $13bn. in foreign debt (out of a total of $20bn. then outstanding). The 'financial rand' (or finrand) was introduced as a dual-exchange rate control, so as to prevent capital flight. This came at a moment when the country's townships were in flames and its factories besieged by militant workers. Relations between Pretoria and international finance were, as a result, contested hotly by the liberation movements, reflecting not only the increased power over South Africa's future wielded by international financiers, but also increased vulnerability (to popular pressure).

The vulnerability was all the more acute because South Africa was suffering an economic slowdown which began during the mid-1970s and became severe during the late 1980s. Particularly in manufacturing, average profitability rates (earnings in relation to capital stock) fell steadily from 40 per cent during the 1950s to less than 15 per cent during the 1980s, and reinvestment dropped by 2 per cent each year during the 1980s. By the trough of the subsequent 1989-93 depression, net fixed capital investment was down to just 1 per cent of GDP, compared with 16 per cent annually during the 1970s.

Dramatic increases in foreign debt were a precursor to the rise of financial speculation. During the late 1970s and early 1980s, South African financiers - especially two inexperienced local-based banks (Nedbank and Volkskas) - had entered the international money markets. Then, between from 1982 and 1984, the Johannesburg Stock Exchange kicked off one of the world's most remarkable bull markets, while simultaneously the capital stock of corporate South Africa stagnated and a deep recession commenced. The shift of flows from productive circuits of capital into financial circuits also coincided with South African corporate borrowers increasing their debt loads.

To exacerbate matters, deregulation of banking began in earnest during the early 1980s. The official De Kock commission recommended lifting prudential requirements and credit and interest rate ceilings, and adopted a 'risk-based' approach to the 'capital adequacy' of a bank, in effect shifting regulation of bank activities from the state to the market. Tax amendments reduced bank tax liabilities to as low as 33 per cent of income by 1993, at a time the corporate tax rate had reached its maximum of 48 per cent. Increased capital flight was facilitated – often illegally (as discussed below) – by financial institutions, and in the period 1985-92 amounted to an estimated 2.8 per cent of GDP in net terms. Indeed, Zav Rustomjee estimated that during the period 1970-88 it was as high as 7 per cent of GDP (Fine and Rustomjee, 1996). This funding, properly directed within South Africa, would have been enough to reverse the period of economic decline to one of marginal growth.

At the same time, total financial sector formal employment nearly doubled, from 106,000 in 1977 to 205,000 in 1996. The most impressive spurts in employment growth were in 1978-85 (from 107,000 workers to 160,000), on the back of the gold boom and associated increase in credit growth, and 1987-90 (from 160,000 workers to 195,000) as deregulation unfolded and a variety of financial innovations were introduced. (Only during the mid-1990s did the banks announce plans to cut jobs – amounting to an estimated 10 per cent of staff – in order to introduce a new wave of labour-saving technology. Branch closures led to more losses.)

The rise of the financial sector was lubricated initially by Pretoria's 'loose money' policies, including the promotion of credit allocation into geographical areas not previously penetrated: black townships (Bond, 2000a). Financial liquidity was growing, with the private sector debt/GDP ratio rising from a stable level of 30 per cent during the post-war era, to 50 per cent during the 1980s and more than 70 per cent by the late 1990s. Housing finance grew especially rapidly during the last half of the 1980s, as banks and building societies invested R10bn. in township bonds. Politically, this addressed an oft-articulated need to identify a new outlet for surplus funds (black townships) which would both enhance the potential for piling on even more consumer credit once collateral (the house) had been established, and introduce an inherently conservatising form of social control (repayment of a twenty-year bond). But the R10bn. was enough to saturate only the top 10 per cent of the market, those who could afford new houses costing in excess of R35,000 (smaller loans were administratively too costly), and it was done in a manner that cemented rather than undermined apartheid urban planning. Moreover, the variable-rate bonds were largely granted at an initial 12.5 per cent interest rate (-7 per cent in real terms).

With an official return to monetarist ideology (as well as anti-apartheid financial sanctions and fear of capital flight), nominal interest rates on housing loans soared to 21 per cent (then 6 per cent in real terms) in 1989, leading to the country's longest-ever depression (1989-93) and, in the process, a 40 per cent default rate on the 200,000 bonds granted to black borrowers. Moreover, the financial explosion also infected commercial real estate and the stock market with untenable speculation. For notwithstanding the overall economic stagnation, between 1982 and 1990 the JSE produced an eight-fold nominal increase in share values, and was the fastest growing stock exchange in the world between 1989 and mid-1992. In 1991 JSE industrial shares increased in price by 56 per cent, while the industrial economy suffered negative growth. Banks, meanwhile, increased their margins between what they charged borrowers and what they rewarded savers (from 2.25 per cent during the late 1980s, the spread doubled by the end of the depression, with a consequent growth in profits to record levels and a huge rise in share values of banking stocks). In short, financial activity borne of economic crisis had helped reshape South African geography, in the process intensifying uneven development.

Meanwhile, capital market funds continued to swell by tens of billions of rands each year, thanks in large part to black worker pension contributions and insurance premiums. These funds were mainly invested in JSE shares and speculative construction of commercial property. In the latter case, overbuilding during the late 1980s quickly generated artificially high land prices in central business districts, and then 20 per cent vacancy rates by 1991 and a full-fledged property market crash. Moreover, the banks began to suffer levels of arrears and defaults within their portfolios comprising company, consumer, housing, white farmer and black taxi loans which were unprecedented in recent history. In particular, they bore huge costs during the crashes of major construction firms and several large conglomerates.

The state's response was to deregulate and privatise even faster, push real interest rates to still higher levels and introduce a regressive value added tax while lowering corporate taxes. Some of these developments initially left South Africa's major banks badly exposed. In 1989, just as the depression began, *The Banker* magazine ranked South Africa's four major banks 290, 303, 412 and 439 in the world in terms of their assets (i.e., the amount of other people's funding they had converted into loans and investments) but just 482, 774, 743 and 540, respectively, in terms of the strength of their underlying capital (the banks' own wealth; the capital/asset ratio is the most common measure of the financial stability of a bank). Yet the return on assets (the most common method of assessing bank profitability) rose most impressively for Standard Bank, for example: from 0.99 in 1991 to 1.09 in 1992, 1.13 in 1993, 1.30 in 1994 and 1.35 in 1995 (far in excess of international norms). First National Bank and Nedcor also steadily increased their earnings during the early 1990s, with only ABSA showing declining (though still substantial) returns on assets during the period. The banks also spent the early 1990s building luxuriant billion-rand headquarters and providing unrivalled compensation packages to senior management.

In short, the banks survived and prospered, gathering record profits (between R750m. and R1.25bn. for each of the four leading banks) during the 1989-93 depression. In past periods, such as the mid-1980s financial panic or the late 1980s disinvestment wave when Standard Chartered and Barclays left, they were bailed out by larger investors: Old Mutual bought Nedbank, the Sanlam-Rembrandt empires supported Volkskas and Bankorp, Liberty Life bought Standard, and Anglo American bought Barclays (renaming it First National Bank). During the early 1990s, new strategies evolved. The most obvious process was intensified concentration within the financial sector. The share of banking sector assets controlled by the four largest banks increased from 69 per cent in 1991 to 84 per cent in 1995. Some of the more sickly banks merged with each other (in the case of Volkskas and Bankorp in the ABSA group). They bought or converted building societies (in the case of Nedbank-Perm, NBS, Saambou and the United-Allied component of ABSA), which had the effect of taking depositors' capital built up over a century in some cases (augmented by generous tax breaks) and effectively privatising it solely on behalf of the present generation. The four large banks rarely engaged in generalised price wars (i.e., the interest rate charged for a loan, or paid to a saver).

Under the circumstances, it was no surprise that bankers were the subject of protests, including by the SA National Civic Organisation which blamed the major banks for:

- 'cementing the geography of apartheid' by financing developers who built housing estates far away from the central cities, often in collusion with corrupt apartheid-era councillors;
- financing 'fly-by-night developers' to put up shoddy housing (90 per cent of township houses had flaws, according to a Housing Consumer Protection Trust report), while bank valuers were sometimes paid to look the other way;
- charging relatively low rates of interest on housing bonds as a baiting technique to solicit marginal borrowers – those affected included

thousands of pensioners and others aged 65 and older who were encouraged to take out bonds so as to construct two rooms and a garage behind their matchbox houses, which in turn were collateralised on top óf the new structure; as interest rates rose marginally, many hundreds of defaults occurred – and then when the interest rate soared by 1989, the banks had nothing to offer those borrowers now unable to pay their bonds;

- failing to develop safety-net mechanisms to protect their own investments and to enable working-class people to retain their homes, as the subsequent recession threw hundreds of thousands of workers out of their jobs, and as rising food prices and new VAT charges ate up monthly incomes;

- not catering for the vast majority of the South African population, both by failing to make available housing loans for less than R35,000 and by not having lending facilities and suitable housing loan products in rural areas where individual title deeds do not exist;

- making loans without adequate buyer education, with no scope for community participation and control, with no forms of civic empowerment, and with no options for co-operatives, land trusts, or housing associations.

In addition to township civics, which carried out 'bond boycotts' in resistance (Bond, 1994), many other South African banking consumers – especially homeowners – complained bitterly about seemingly-capricious interest rate increases; a limited range of financial products; a legacy of poorly-designed low-cost lending initiatives; a wave of foreclosures and evictions; the 1995 decision to raise interest rates for low-income homeowners (at the same time lower rates were being offered to professionals); the 1995-98 spate of bank branch closures and the simultaneous disowning of more than a million low-profit savings accounts; and systematic 'redlining' (geographical discrimination) of black and desegregating urban neighbourhoods. In addition, consumer advice offices received more complaints about financial institutions than about any other sector of commerce. General consumer concerns about bank practices included misleading advertising, inadequate consumer education, lack of competition between banks, rising fees for financial transactions (even at Automated Teller Machines), lack of access to bank facilities for township and rural financial services consumers, and lack of responsiveness to complaints. Moreover, because banks largely failed to serve South Africa's lower-income consumers, there was a notable rise in loan sharks (*mashonisas*), fraudulent Ponzi schemes and other exploitative financiers (including touted NGO creditors) in township and rural markets. Credit available only at interest rates of 350 per cent per annum was not at all unusual for informal-sector borrowers with no credit history. Notwithstanding a continual barrage of rhetoric about South Africa's poor savings rates, banks actively discouraged low- and moderate-income people from opening accounts.

The cost of credit was also a major deterrent to consumers. Bankers regularly violated the Usury Act on overdraft facilities, which generated a new cottage industry of small-time accountants calculating interest for angry bank clients. ABSA alone allegedly overcharged customers by R1bn. The vast number of publicised consumer complaints against banks suggested a stunning level of officially condoned fraud, emblematised by the fact that the bureaucrat responsible for Usury Act enforcement was downgraded and ultimately dismissed. More generally, thanks to the Reserve Bank's post-1989 tight-money policy, which was ratcheted even tighter in 1995, real interest rates on South African government bonds had reached more than 10 per cent by the mid-1990s and 14 per cent by mid-1998, compared with less than 5 per cent in Britain and Germany, and approximately 3 per cent in the US, Japan and Australia. Such historically unprecedented real interest rates generated an enormous inflow of footloose foreign finance to South Africa during 1995, following the lifting of the finrand in March. It was not only the bond markets that prospered from the inflow. More than half the turnover on the Johannesburg Stock Exchange during 1995 was of foreign origin. But the hot money also soon wreaked havoc on the currency, for it drained out just as quickly when negative herd instincts emerged a year later, leading to a crash in the currency during the first months of 1996, and again in 1998 and 2000-01.

These outcomes represented untenable trends combining ascendant finance and worsening uneven development. What post-apartheid financial regulatory provisions were mandated, and what was accomplished during the ANC's first term in office?

Post-apartheid Domestic Financial (De)Regulation and Crisis

The first democratic South African government adopted an extremely ambitious development policy framework, established through priorities identified by mass social and civic movements during struggles of the 1980s and early 1990s. The *Reconstruction and Development Programme* (RDP) of 1994 includes expansive promises regarding bank regulation and development finance. In general, however, the RDP failed to meet expectations (Bond and Khosa, 1999; Bond, 2000b, chapter 3). In particular, the failure to establish a strong financial regulatory system reflects the balance of forces in the key economic ministries.

Part of the problem lay in the economic legacy of apartheid. The ANC government inherited a long-term structural crisis. Problems included persistent overcapacity and overproduction of (relatively uncompetitive) luxury manufactured goods for the (mainly white) upper-income consumer market, side-by-side with growing surpluses of unemployed black workers, heightened financial speculation and intensifying geographical unevenness. The top 5 per cent of South Africa's population consumed more than the bottom 85 per cent resulting in a Gini coefficient (the main measure of income disparity) that matched Brazil and Nigeria as major countries with the worst levels of inequality; these statistics only worsened during the 1990s.

While the RDP called for accountability, transparency and extension of financial services to those who presently do not have access to these services, progress on these objectives was minimal and even negative. The RDP suggested a variety of anti-discriminatory measures for the financial sector, covering race, gender, location and other non-economic factors, which were ignored by the ANC government. The RDP mandate to consider removing the Usury Act exemption was contradicted by the Department of Trade and Industry, which liberalised virtually all forms of lending and raised the interest ceiling. And while the RDP envisaged a 'Housing Bank' backed by a Guarantee Fund, subsidised so as to lower the interest rate for low-income borrowers, the Housing Ministry raised the rate by 5 per cent and relied on commercial banks, to its regret. The RDP suggested that the state 'encourage community banking', by changing regulations (while protecting consumers) and placing accounts in community banks, but when the main community bank was in financial trouble in 1996, the Finance Ministry let it collapse. The RDP called for pension/mutual funds to be more accountable to members, and for Mutual Funds boards to be transformed, but the Finance Ministry instead allowed the two main mutual insurance firms (Old Mutual and Sanlam) to become listed institutions on the stock market, wiping out generations of mutual equity investments in favour of (white dominated) account holders. And while the RDP called on government and the Reserve Bank to intervene more actively against 'illegal capital flight' than earlier under existing exchange controls, the Bank and Finance Ministry instead merely abolished the finrand.

Thus financial sector laxity became the rule. Following Gencor's purchase of Shell Oil's Billiton mining group, most of the country's largest firms – Anglo American Corporation, Liberty Life, SA Breweries, Old Mutual and Didata – announced the transfers of their primary financial headquarters to London, and De Beers delisted entirely, with expectations of its likely relocation to London. Mining houses were another vehicle for disinvestment during most of the 1990s, with widespread allegations of massive transfer pricing to shell companies in places like Zug, Switzerland. From around 1991, the four major banks opened more branches in the Cayman Islands, Panama, the Isle of Man, the Bahamas, Guernsey, Jersey, Zurich and other hot money centres than in all South Africa's black townships combined (in fact, the banks closed most of their township branches after a housing finance crisis emerged in 1990).

Also crucial to the flow of funds out of South Africa was the willingness of the Reserve Bank to accept responsibility for banking and corporate exposure in international markets. This took several forms, including an enormous 'lifeboat' for Bankorp when it began to fail in 1985; it was transferred to ABSA and increased to R1.125bn. – as a 'loan' to be repaid at 1 per cent interest - in part to help ABSA wind up extremely poor quality assets tied to corrupt banking practice. There was also Reserve Bank 'forward cover' protection, namely billions of rands in subsidies against the risk of currency devaluation – based on more than R100bn. of foreign loans by the late 1990s – essentially donated to those firms which could borrow internationally, notwithstanding the enormous cost involved given the sliding value of the rand. The initial rationale was that due to financial sanctions, the Reserve Bank continually needed to acquire foreign reserves.

Finally, outright corruption also played its part in South Africa's reintegration into the international markets. An unending stream of financiers from major banks, insurance houses and stockbroking firms were unveiled, but this was probably the tip of the iceberg. The Durban director of the Reserve Bank, for example, was convicted in 1992 of having stolen $1bn. through foreign exchange fraud. In 1994 the Witwatersrand Attorney General, investigating late-1980s foreign exchange activities of South African banks, issued a report alleging fraud in every major transaction approved by banks and the Reserve Bank. South Africa was cited by Britain's Centre for International Documentation on Organised and Economic Crime as 'a prime target for future growth in international economic crime and money laundering' (Bond, 2000b, chapter 1).

Deregulation of local financial markets limited South Africa's capacity to withstand subsequent economic crises (Bond, 2001). As a result of raiding by speculators, the currency crashed by 30 per cent over a matter of a few weeks in early 1996 and mid-1998, and again by more than 50 per cent over an 18-month period during 2000-01. In addition to large players in New York and London, South Africa's own banks were very active in betting against ('shorting') the rand, and recorded large foreign-exchange account profits as a result. Standard Bank's forex-trading rank rose from 96th in the world in 1998 to 65th a year later, while First National Bank's rose from 111th to 78th. In addition, the two most active currency speculators in the world (each with more than 7 per cent market share), Citibank/Salomon Smith Barney and Deutsche Bank, both have active Johannesburg offices.

Even the founder of Liberty Life insurance, Donald Gordon, expressed extreme frustration in 1999, after losing enormous amounts of paper value to speculators fleeing Liberty's $350m. 'euro-convertible bond' issue, in the process crashing the insurance company's share value. Gordon remarked ruefully (*Sunday Independent*, 16 May 1999), 'In the name of short-term gain for a few, these people have been allowed to undermine most of the emerging markets. In South Africa [permitting foreign speculation on local assets] was the financial equivalent of allowing hostile war boats free rein along our coast. It is a destructive activity that undermines the very core of our sovereignty'.

Yet there are still more avenues available for deregulation, as well as mergers and acquisitions across the financial sector. Due in part to the highly concentrated, uncompetitive character of South African financial markets, and the relatively low-level development of consumer awareness, innovation has thus far not gone as far as in other financial markets. Major derivative instruments associated with currency speculation such as spots, swaps, futures, and forwards are not significant in South Africa's international financial relations, compared with simple stock, bond and currency purchases/sales. But the roles of financial derivatives will no doubt increase, regardless of how convertible they become internationally. The South African Futures Exchange has taken nearly a decade to get off the ground, and is unpopular because of high commission fees, but remains a potential player. Other kinds of derivatives are increasing, and some are being marketed internationally. For example, the World Bank's International Finance Corporation promotes the 'Gateway' housing bond securitisation for high-income households' debt, not for the millions of low-income South Africans who require additional affordable mortgage funds. The scheme, in any case, is modelled along the lines of US housing bond securities which proved terribly dangerous to Savings and Loan Associations during the late 1980s.

All of these factors make South Africa more vulnerable to systematic international financial risk from institutions which, as Soros noted, are 'ill-suited to allocate international credit'. And yet Finance Minister Trevor Manuel announced to the US-South Africa Business Council in 1999, that 'South Africa remains committed to the gradual liberalisation of the capital account. These controls will continue to be reduced in a manner that does not destabilise the market, while ensuring that the financial system manages its risk exposure in a prudent manner'.

Conclusion: Movements for Democratic Finance?

The degeneration of South Africa's economy, and the extreme volatility of currency and financial markets notwithstanding low state deficits and a persistent trade surplus, finally brought home the need to tighten financial controls. Probably most important was the massive outflow of profits and dividends to the new London financial headquarters of the largest South African corporations in 2000-01, which compelled Trevor Manuel to reregulate financial markets slightly in the 2001-02 budget, by prohibiting the 15 per cent offshore investment allowance for major institutions. Reserve Bank Governor Tito Mboweni railed against both global financial speculation on the rand, and local denial of credit to desegregating neighbourhoods. Even Treasury Director-General Maria Ramos (2000, p. 126) conceded that 'in an extreme crisis situation, default will have to be part of the equation. In that case, the only way in which you are going to prevent a short-term outflow of capital is through some pretty tough exchange control measures. I don't know if there are too many options available'. Would a top-down reform of the financial system be feasible, given the apparent government mood shift at the outset of the 21st century?

Other contradictions were also emerging, not least in international financial relations. Countries as diverse as Argentina and Zimbabwe were falling into formal debt default. Thabo Mbeki's New Partnership for African Development pledged to demand higher levels of African debt cancellation by international lenders (Nelson Mandela had cancelled Namibia's debt in 1996). Yet critics wondered how the new approach squared with South African officials':

- own agreement to repay the apartheid foreign debt, and repeated claim that there *was no* foreign debt owed by South Africa at the time of liberation (by ignoring roughly $25bn. parastatal and private sector debt, for which the South African state inherited repayment and guarantor responsibilities);
- refusal to demand reparations for previous foreign credits to the apartheid regime (e.g., at the September 2001 World Conference Against Racism);
- endorsement, repeatedly, of the Highly Indebted Poor Countries initiative of the G-8, IMF and World Bank, which proved to be a severe distraction from the cause of debt cancellation;

- failure to cancel South African loans to Mozambique, including 1990s credits to repair damage done to Cahora Bassa electricity transmission lines by apartheid-backed Renamo and to resettle dissident right-wing Afrikaner farmers.

If not from an erratic officialdom, which appeared to maintain neo-liberal inclinations – e.g. in the New Partnership for African Development, home-grown austerity, and most ongoing international relations relating to trade, foreign investment and debt – where then might the catalyst to financial system reform be located? Jubilee South Africa, the Congress of South African Trade Unions, the South African Communist Party, the South African Non-Governmental Organisation Coalition and other far-sighted civil society groups soon began motivating for a different distribution of financial sector resources. At its August 1999 national congress, for example, the Congress of South African Trade Unions affirmed the following position:

1. Government has gradually lifted and is still committed to the further lifting of exchange controls. The fundamental problem with lifting exchange controls is that it led to increased mobility of speculative capital in and out of the country, and to domestic capital flight. Such flights have destabilising effects on the economy. The free movement of short-term speculative capital exacerbates economic instability. Much of the foreign investment realised over the past few years consists mostly of speculative flows. Apart from increased instability, exchange control liberalisation is also likely to reduce rather than increase the resources available for productive domestic investment. Furthermore, at a broader political economic level, weakened exchange controls make South Africa more vulnerable to international markets and hence constrain our domestic policy options.
2. Government should increase its regulation of financial markets significantly. Exchange controls should be maintained and further strengthened in order to protect the domestic economy from the flight of capital and guarantee the efficacy of government intervention in the economy. Specific taxes to reduce the attraction of speculative investment relative to productive investment should be introduced, through a system of incentives and disincentives. An alternative approach to the regulation of financial markets would be to redirect investment streams much more.

Unquestionably, there are several global and local challenges for the South African government, to follow its strategic self-interest in relation to financial markets:

- halting further financial liberalisation and reconsidering those aspects of recent liberalisation which are economically and morally questionable;
- imposing feasible exchange controls at the earliest opportunity, and certainly prior to a renewed state of financial turmoil and crisis when they might become vitally necessary;
- restoring confidence in the Reserve Bank, as a crucial logistical vehicle for exchange controls, at a time of new Bank leadership and systems;
- distinguishing between future inflows of hot money (which should be actively discouraged) and future production-oriented FDI;
- revisiting the structure of the current and capital accounts, including imports and the inherited apartheid-era foreign liability structure;
- lowering interest rates;
- redirecting financial resources into productive purposes, including meeting human needs, away from largely speculative and unproductive outlets.

Notwithstanding occasionally ambitious South African government rhetoric, it is only through active grassroots-based advocacy that such domestic – as well as global-scale – financial sector reforms can be catalysed and won, so as to transform the geography of financial markets in the interests of society. A top-down reformer, John Maynard Keynes (1933, p. 769), once remarked:

> I sympathise with those who would minimise, rather than with those who would maximise, economic entanglement among nations. Ideas, knowledge, science, hospitality, travel – these are the things which should of their nature be international. But let goods be homespun whenever it is reasonably and conveniently possible and, above all, let finance be primarily national.

Keynes was correct about the need to discipline finance in its domestic environment, first and foremost. South Africa's post-apartheid rulers have been mainly doing the opposite, leaving serious reform advocacy to civil society movements. The latter are inexorably drawn to the conclusion that their struggle for social justice locally is bound up in the fight against the 'global apartheid' that has been exacerbated by the extreme uneven development of finance across time and space these past two decades. It is only by assuring secure and more equitable domestic financial relations in countries like South Africa, in the context of turbulent international flows that suggest policies of financial delinking, that a solid foundation can be build for a new, sustainable domestic *and global* financial architecture.

References

Ally, R. (1994), *Gold and Empire*, University of the Witwatersrand Press, Johannesburg.

Bank for International Settlements (1999), *Central Bank Survey of Foreign Exchange and Derivative Market Activity, 1998*, Basle, Switzerland.

Barclays Bank (1938), *A Banking Centenary*, Barclays, London.

Bond, P. (1991), *Commanding Heights and Community Control: New Economics for a New South Africa*, Ravan, Johannesburg.

Bond, P. (1994), 'Money, Power and Social Movements: The Contested Geography of Finance in Southern Africa', in S. Corbridge, R. Martin and N. Thrift (eds), *Money, Power and Space*, Basil Blackwell, Oxford.

Bond, P. (1998), 'A History of Finance and Uneven Geographical Development in South Africa', *South African Geographical Journal*, vol. 80, pp. 23-32.

Bond, P. (1999a), 'Uneven Development', in P. O'Hara (ed), *The Encyclopaedia of Political Economy*, Routledge, London, pp. 1198-2000.

Bond, P. (1999b) 'Finance Capital', in P. O'Hara (ed), *The Encyclopaedia of Political Economy*, Routledge, London, pp. 345-47.

Bond, P. (2000a), *Cities of Gold, Townships of Coal: Essays on South Africa's New Urban Crisis*, Africa World Press, Trenton.

Bond, P. (2000b), *Elite Transition: From Apartheid to Neoliberalism in South Africa*, Pluto Press, London, and University of Natal Press, Pietermaritzburg.

Bond, P. (2001), *Against Global Apartheid: South Africa meets the World Bank, IMF and International Finance*, University of Cape Town Press, Cape Town.

Bond, P. and Khosa, M. (eds) (1999), *An RDP Policy Audit*, Human Sciences Research Council, Pretoria.

Fine, B. and Rustomjee, Z. (1996), *The Political Economy of South Africa*, Witwatersrand University Press, Johannesburg, and Christopher Hurst, London.

Frank, A.G. (1967), *Capitalism and Underdevelopment in Latin America*, New York, Monthly Review.

Gelb, S. (1989), 'The Origins of the South African Reserve Bank, 1914-1920', in A. Mabin (ed), *Organisation and Economic Change*, Ravan, Johannesburg, 48-77.

Giliomee, H. (1989), 'Aspects of the Rise of Afrikaner Capital and Afrikaner Nationalism in the Western Cape, 1870-1915', in W. G. James and M. Simons (eds), *The Angry Divide: Social and Economic History of the Western Cape*, David Philip, Cape Town.

Harvey, D. (1989), *The Condition of Post-Modernity*, Basil Blackwell, Oxford.

Henry, J.A. (1963), *The First Hundred Years of the Standard Bank*, Oxford University Press, London.

Henwood, D. (1998), *Wall Street*, Verso, London.

Innes, D. (1984), *Anglo American and the Rise of Modern South Africa*, New York, Monthly Review.

Keegan, T. (1986), *Rural Transformations in Industrialising South Africa: The Southern Highveld to 1914*, Ravan, Johannesburg.

Keynes, J.M. (1933), 'National Self-Sufficiency', *Yale Review*, vol. 22, pp. 755-69.

Mabin, A. (1984), *The Making of Colonial Capitalism: Intensification of the Economy of the Cape Colony, South Africa, 1854-1899*, unpublished PhD thesis, Simon Fraser University, Vancouver.

Mabin, A. (1985), 'Concentration and Dispersion in the Banking System of the Cape Colony, 1837-1900', *South African Geographical Journal*, vol. 67, pp. 141-59.

Mabin, A. (1986), 'The Rise and Decline of Port Elizabeth, 1850-1900', *International Journal of African Historical Studies*, vol. 19, pp. 275-303.

Mabin, A. (1989), 'Waiting for Something to Turn Up? The Cape Colony in the 1880s', in A. Mabin (ed), *Organisation and Economic Change*, Ravan, Johannesburg, pp. 21-47.

Manuel, T. (1999), Address by the Minister of Finance, Trevor Manuel, to the US-South Africa Business and Finance Forum, 24 September.

Nattrass, J. (1981), *The South African Economy*, Oxford University Press, Cape Town.

O'Meara, D. (1983), *Volkskapitalism*, Cambridge University Press, Cambridge.

Phimister, I. (1992), 'Unscrambling the Scramble: Africa's Partition Reconsidered', paper presented to the African Studies Institute, University of the Witwatersrand, Johannesburg, 17 August.

Ramos, M. (2000), 'Comment on A New Framework for Private Sector Involvement in Crisis Prevention and Crisis Management by Jack Boorman and Mark Allen', in J. Teunissen (ed), *Reforming the International Financial System: Crisis Prevention and Response*, Forum on Debt and Development, the Hague, pp. 125-27.

Schumann, C.G.W. (1938), *Structural Changes and Business Cycles in South Africa, 1806-1936*, Staples, London.

Soros, G. (1997), 'Avoiding a Global Breakdown', *Financial Times*, 31 December.

7 South Africa's Information Technology Economy: Directions and Key Policy Issues

CHRISTIAN M. ROGERSON

Introduction

The period from the late 1960s to the late 1980s witnessed the convergence of a series of scientific and technological innovations to constitute what Manuel Castells (1996, 1998) refers to as 'a new technological paradigm' and birth of a global informational economy. Within this new global economy, Africa often is seen as largely excluded or, at best, is marginalised (Castells, 1998). Indeed, Castells (1996, p. 136) goes so far as to observe that 'most of Africa ceased to exist as an economically viable entity in the informational/global economy' and Mabogunje (2000, p. 169) writes of the Africa 'dis-connect'. Nevertheless, across the African continent the position of South Africa is viewed as something of a special case because of the country's strong technological linkages in the global economy (Castells, 1998). Although Gillwald (2000, p. 4) argues 'South Africa appears to be struggling to transform itself into an informational society despite considerable political will', as measured by all available indicators regarding the new informational economy, it is evident that South Africa has established firmly a leadership role within Africa (James, 1999; Creamer, 2001). Moreover, President Mbeki has made several statements that the development of telecommunications and information technology is a major priority for South Africa within the Millennium Africa Recovery Plan for eradicating poverty and nurturing sustainable growth and development (Erwin, 2001).

This chapter seeks to map out the salient features of South Africa's information technology (IT) economy and analyse the major contemporary policy issues and challenges concerning its wider development. Existing source material on South Africa's IT economy, however, is relatively undeveloped. The most important available material on the IT economy currently emanates from the outputs of the South African Information Technology Industry Project (SAITIS) which is housed within the Department of Trade and Industry (James, 1999, 2000; SAITIS-DTI, 2000; James et al., 2001). This project was initiated during 1994-95 with the objective of bridging the global and local development gaps in the use and application of information and communication technologies (James et al., 2001). Within this particular project, a broad perspective is taken on defining the IT economy by using the terminology of the 'ICT sector'. Following definitions applied by the OECD, this ICT sector is acknowledged to include, *inter alia*, the manufacture of computer hardware and telecommunications equipment and services such as IT professional services (custom software application development and maintenance), computer software (packaged software products), and telecommunications service (SAITIS-DTI, 2000, p. 5).

South Africa's IT Economy – Development, Structure, Geography

The IT economy in South Africa has burgeoned from humble beginnings to a size estimated at $6.4bn. by 1992 and $9.7bn. by 1997. For the last five years at least, the South African IT economy has been on a high growth path with real growth rates of 10 percent or more per annum (Franz, 2000). This surge in growth is viewed as a mirror of international trends as well as the result of local restructuring following the 1994 elections. Globally, South Africa ranks the twentieth largest country market for ICT products and services, accounting for 0.6 percent of worldwide revenues (SAITIS-DTI, 2000, p. 9). As compared with other developing countries, the rate of adoption of e-commerce by South Africans shows a relatively high growth rate (Republic of South Africa, 2000, p. 68) and considerable potential is seen in the future growth of different types of e-commerce. According to Moodley et al. (2001) e-commerce in South Africa is dominated by business-to-business (B2B) rather than the much-hyped business-to-consumer transactions (B2C). Overall, B2B trading increased from a mere R15m. in 1997 to an impressive R3.9bn. by 1999 (Moodley et al., 2001, p. 2). Further rapid expansion of B2B transactions is envisioned by industry leaders with B2C transactions growing only later as a 'natural progression' (Ord, 2001).

Historically, South Africa's IT economy evolved during the 1960s and 1970s as a result of technological transfers from the United States and the United Kingdom and the subsequent catalytic effect of this transfer on education and skills (James, 2000, p. 6). At its inception, during the 1960s and 1970s the market offered limited choice and was characterised by the struggle for market share that occurred between IBM and 'The Bunch', which comprised of Burroughs, Univac, NCR, CDC and Honeywell. Initially, it is argued there existed a clear separation of the IT and communications industries. Subsequently this divide became so blurred that the terminology of the ICT sector is viewed as most appropriate to describe this evolving economy (James, 1999, p. 65). Nevertheless, one continuing thread is the high level of dominance of South Africa's IT economy by foreign multinational enterprises, such as Unisys, Microsoft, Intel, Dell or Novell, through their local subsidiaries (James, 2000, p. 7). In terms of listed IT companies on the Johannesburg Stock Exchange, the major players are Dimension Data, Comparex, Datatec and the Altron Group. Overall, it is argued that like most developing countries, South Africa is 'a better user than designer of complex technology products' with most local success stories, particularly in software development, 'limited to reworking or local integration of internationally designed packages' (James, 1999, p. 87).

The most notable exception in the IT economy is the growth of Dimension Data which was formed in 1983 in response to demands from the South African military apparatus. The enterprise has grown rapidly from a networking hardware reseller to a network integrator and e-business enabler, with a near global footprint (Bidoli, 2000). In terms of market capitalisation, this South African multinational, at one time, was ranked as the largest IT company on the London Stock Exchange. The operations of this company are highly globalised; with operations in over thirty countries and in 2001 DiData was pursuing an aggressive strategy of corporate expansion both in the USA (Bidoli, 2000) and the Netherlands (Stones, 2001).

Finally, in terms of the ownership structure of the IT economy note must be made of several initiatives to support black economic empowerment (BEE) companies in the IT economy. For example, in the telecommunications sector there are commitments made by national government to support SMMEs and greater ownership and control by persons from historically disadvantaged groups (Hodge, 2000). These BEE enterprises which were initially so defined if they were 50 per cent under black ownership and later when they fulfilled the additional criteria of having a roughly 30 per cent black skills base, are nurtured by contracts awarded by South Africa's major parastatals. It is argued that many of these BEE enterprises are developing to the extent that they may emerge within a five-year horizon into efficient private sector competitors (Spicer, 2001).

The location of South Africa's IT economy has been established using data derived from the Matrix Marketing database (Rogerson, 2001). The findings from this investigation disclosed the existence in 1999 of a total of 1,220 IT service enterprises with an estimated 64,109 employment opportunities. Within the various categories of IT service activities both the largest numbers of enterprises and employment opportunities are in the activities of distributors, systems integrators and specialist retailers of IT equipment and supplies. Taken together, these categories account for 66 per cent of enterprises and 76 per cent of total employment opportunities within the IT service economy. The largest number of enterprises are found as systems integrators whereas distributors account for the largest individual element of employment.

Analysis of the location of the information-technology service economy discloses a remarkable degree of spatial clustering or concentration of IT activity. Of the national total of IT jobs, 78 percent are clustered in Gauteng. The intense spatial agglomeration of the IT service economy is indicated by the fact that 70 percent of national IT employment (nearly 45,000 jobs) is concentrated in the Greater Johannesburg Metropolitan Area. By contrast, in terms of employment Cape Town (with 5,000 jobs) is recorded with an 8 per cent share of national IT employment. In terms of numbers of IT enterprises a similar geographical pattern emerges with Gauteng accounting for 68 per cent of total enterprises.

Within all the major segments of the IT service economy, the dominance of Gauteng is once again in evidence. In terms of national employment opportunities the share of Gauteng was as follows; systems integrators (82 per cent), distributors (95 per cent), netware specialists (91 per cent), software development (93 per cent) and internet service providers (68 per cent). Outside Gauteng, local specialisations appear to exist only in the Western Cape for internet development and internet service provision. At a fine-grained level of analysis, the Greater Johannesburg Metropolitan area is by far the leading focus for most aspects of the IT service economy. The only notable exceptions are the relative importance of Cape Town in internet services (second to Johannesburg) and the leadership of the highly specialised area of internet development by enterprises based in Pretoria and Cape Town-Stellenbosch (Rogerson, 2001).

In terms of the factors that influenced the decisions to locate IT businesses, the core issues relate to those of market considerations and the location of the information-intensive users (Rogerson, 2001). It is estimated that 72 per cent of all national IT expenditure occurs in Gauteng province (Dorfling, 1999). The locational choice for 95 per cent of Gauteng IT enterprises is influenced by the province's status as IT and industrial hub of the country. Other issues that are important relate to availability of skilled personnel as a secondary factor for business location in Gauteng (James, 1999). Overall, Gauteng was rated as the best potential business location by a national sample of forty IT service enterprises. The only perceived advantages of the Western Cape (as expressed by Gauteng enterprises) relate to the region's pleasant residential environment. This factor was, significantly, often reinforced by the fact that the Western Cape was the original place of residence and/or business of the owners of IT service enterprises in Gauteng.

Support Initiatives - National and Local

At national level, several initiatives have been launched to support further the economic health of the IT economy. Broadly speaking, these initiatives fall into two groups. First, a suite of programmes relate to either (1) enhancing the human resource or skills base for the IT industry, or (2) supporting the development of a network society, including infrastructure, that would incorporate also South Africa's previously disadvantaged and marginalised communities. Under this rubric there are encompassed several programmes which are operated by different national government departments, parastastals, the private sector and NGOs. The Departments of Education and Communications have been actively engaged in several training and education initiatives that are targeted to support the production of a greater flow of IT professionals through schools, technikons and universities. In addition, there are a range of IT-related interventions which are geared to establish a networked society 'that empowers people in the way they work, live and play, and to make South Africa globally competitive' (James, 1999, p. 39). Improvements in the creation of a physical infrastructure for access to information are of central importance and include for example the initiation of community information centres designed to increase the access of previously disadvantaged people in townships and rural areas to information services.

A second category of national level initiatives for supporting the IT industry relate more firmly to expanding the competitiveness of existing enterprises as well as fostering new IT enterprise development, especially in terms of small, medium and micro enterprise (SMME) development. The SAITIS study identifies a host of programmes operated by the Department of Trade and Industry which are designed to stimulate economic growth and are potentially applicable to the IT economy (James, 1999). Among the most notable of these are the competitiveness fund, the sector partnership fund and a range of industrial investment incentives. In addition, the DTI's Technology and Human Resources for Industry Programme (THRIP) supporting the development of technology and of skills training to improve South Africa's competitiveness and the Innovation Fund, a programme of the Department of Arts, Culture, Science and Technology, are two further support structures which can be accessed for nurturing the IT economy. Beyond national initiatives, the growth potential of the IT economy in South Africa has been recognised by local economic development planners. Aggressive competition is emerging between the Gauteng and Western Cape provinces in terms of seeking to attract the largest share of IT enterprises and bidding to become the 'Silicon Valleys of Southern Africa' (Addison, 2000, p. 23). In Gauteng, this IT promotion is focussed around the Spatial Development Initiative and the planning of the province as South Africa's 'smart region'. As the planning for Gauteng's SDI has evolved, the vision has consolidated into developing the province into a 'smart hub' (Gauteng Province, 1997), which would be focussed upon knowledge-based economic activities. Within the planning for Gauteng as a 'smart hub' are several projects. The IT sector is a focus for investment promotion at the planned Industrial Development Zone at Johannesburg International Airport. Another notable project is an 'innovation hub' as the axis for a development corridor running along the highway that links Johannesburg and Pretoria. It is recognised that of critical importance in this innovation corridor are 'the regional economy of creative industries, innovation, IT and telecommunications "connectivity" into global information networks and markets' (DTI, 1998, p. 4). During February 2000 an initiative was established which serves to link a set of business, education and research institutions in the Pretoria area in order to create, nurture and grow technology-led or high-growth knowledge-based businesses. Long-term planning involves developing at a new greenfield site a high technology industrial park with high bandwidth connectivity through fibre optic cables, an incubator and innovation centre, a venture capital company and an investor information and support system (Dorfling, 1999, p. 2). Overall, the projects coming on stream as part of the Gauteng SDI are geared to strengthen the province's competitiveness for both high technology manufacturing and information service activities (Rogerson, 2000, p. 337).

In the Western Cape there are several initiatives designed to boost the province's competitiveness for knowledge-based activities. The local development agency prioritises IT activities as one of the 'core clusters' for development in the Western Cape (City of Cape Town, 2001). Accordingly, there is extensive place marketing of the Western Cape as South Africa's technology hub, using the backdrop of Cape Town's scenic Table Mountain as the basis for asserting the region's high quality of life and environmental attractions. Cape Town is marketed as 'a city geared to meet the challenges of the knowledge economy'. The Cape Information Technology Initiative (CITI) is seen as potentially a model for regional high technology cluster initiatives in South Africa (James, 1999, p. 52). The CITI is an independent Section 21 company that seeks to catalyse the development of a dynamic cluster of IT industries in the Western Cape. The goal is to establish the Western Cape as the IT 'Gateway to Africa'. The initiative is grounded in the current theory and practice of cluster development and arose out of discussions held in 1997 between the private sector, local universities and provincial government. The CITI is a public-private partnership that is housed in a business incubator on Cape Town's foreshore and aims to nurture new IT companies and develop an IT cluster in the province (Bisseker, 2001). In addition, during 1999 it was announced that CITI had secured seed funding for developing a world-class Web site, which is seen as pivotal for galvanising the IT cluster in the Western Cape. Another high-profile initiative to support the development of the Western Cape as a competitive base for knowledge-based enterprises is the development of South Africa's largest science or technology park. The Capricorn Technology and Industrial Park in Cape Town is marketed as the 'Western Cape's technology innovation hub', a 'world-class location' which provides 'an opportunity for companies to locate their manufacturing, office, research and distribution activities on one fully integrated site'.

Major Policy Challenges for IT development

The IT sector has been identified as one of the potential 'high growth areas' which are to be stimulated by specific national government intervention (Dobson, 2001). Nevertheless, a number of challenges need to be addressed for the continuing growth of South Africa's IT economy (James et al., 2001). First, it is argued that in South Africa, 'the productivity potential of the digital economy is constrained by a lack of critical infrastructure' (Moodley et al., 2001, p. 3). More particularly, a major priority for national government is expansion and upgrading of the electricity grid and of telecommunications infrastructure in order for a roll-out of broadband infrastructure. Second, in disadvantaged areas the reduction of the digital divide between the information rich and information poor necessitates consideration of public subsidies and tax concessions in order to finance the construction of high-speed internet networks (Republic of South Africa, 2000; Moodley et al., 2001). In terms of disadvantaged areas, Stavrou et al. (2000, p. 12) conclude that, for the immediate future, e-commerce 'will have little impact on the people in greatest poverty', albeit it can be a tool to support SMME development if other issues such as training, credit and general business support are also dealt with. The possibilities for the ICT sector to be a basis for fostering 'virtual agglomerations' in support of rural SMMEs are highlighted by Bourgouin (2002).

Third, there is a need to liberalise the telecommunications industry, whilst also ensuring that a strong regulator with a clear mandate is in place

(James, 1999, 2000). The so-termed 'managed liberalisation' of South Africa's telecommunications industry reached a new phase in May 2002 when limited competition to Telkom was introduced in the form of a second network operator which is expected to bring in significant new investment (Hodge, 2000; Creamer, 2001). The danger of a duopoly in telecommunications is prompting consideration of granting a third national network operator so providing greater competition (Campbell, 2001). Fourth, a clear broader need is recognised to develop a stronger entrepreneurial base in the IT economy (James, 2000). Issues which are critical relate to improving access to venture capital, increasing incentives to make the IT economy more attractive to local and foreign investors, entrepreneurial training and developing a facilitative regulatory and policy environment to support such development (SAITIS-DTI, 2000).

Fifth, it is suggested that a re-examination is required of existing DTI policies pertaining to the IT economy. More especially, policy interventions designed '*inter alia*, to promote technology transfer for businesses interested in adopting Internet technology, providing small business technical support, facilitating export development and promotion (via e-commerce) and facilitating the diffusion of e-business best practices in the private sector, are pressing priorities' (Moodley et al., 2000, p. 4). Existing barriers to diffusion that need to be addressed are access costs, network capacity and the speed of data transmission and low Internet use by many enterprises. Overall, it has been observed that the South African government 'needs to make substantial investments in rule making' in terms of a regulatory environment to facilitate the IT economy. An important step in this process was the preparation by the Department of Communications of the Green Paper on E-Commerce which appeared in November 2000 (Republic of South Africa, 2000).

Finally, the area that is perhaps of most critical policy significance relates to skills development for the IT economy (James et al., 2001). Human resource development and the emigration of skilled personnel or the 'brain drain' has attracted much recent public attention and detailed research (Crush et al., 2000; McDonald and Crush, 2000). Human resource or skills development requires building upon the existing education and training systems and institutions in order closely to harmonise their outputs to the IT needs of the country (SAITIS-DTI, 2000). Although the actual extent of 'brain drain' is imprecise, recent work confirms that 'the South African brain drain is clearly much more significant than the official figures suggest' (Brown et al., 2000, p. 46). Another study asserts that 'official figures drastically underestimate the extent of skills out-migration' (Mattes and Richmond, 2000, p. 10). In addition, what is termed 'brain drain pessimism' has centred 'on the exit of skills from South Africa with little focus on what South Africa can do to attract more skilled workers from abroad or to keep those who stay here' (Mattes et al., 2000, p. 29). Post-1994 South African immigration policy has 'placed a very low premium on skills import from anywhere' (McDonald and Crush, 2000, p. 4). Since 1994 there has occurred a consistent decline in terms of the temporary and permanent immigration of skilled immigrants and national government 'has come only reluctantly to the view that skills immigration is not necessarily disadvantageous to South Africans' (ibid., 2000, p. 6).

Whilst certain sectors of the South African economy so far have been little impacted by brain drain, the IT economy is the most outstanding example of a sector in which professional skills are now in short supply (James, 1999; SAITIS-DTI, 2000; James et al., 2001). This shortage of IT skills is partly a consequence of the lack of co-ordination between the education system and the labour market and of demand outstripping supply. Equally important the shortage of IT skills is linked to both the emigration of skilled personnel and paralysis of government immigration policies concerning 'brain gain' (Crush et al., 2000; McDonald and Crush, 2000; Mattes et al., 2000; Rogerson and Rogerson, 2000). Issues of crime and safety, levels of taxation, and of international remuneration levels emerge as the underlying factors behind the brain drain for IT personnel. As was argued in the most detailed investigation so far undertaken, 'a global labour market has emerged in this sphere, and South Africa is rapidly losing competitive ground to the aggressive recruitment strategies and lucrative employment packages of enterprises in the USA, Western Europe and Australasia' (Rogerson and Rogerson, 2000, p. 36). Moreover, the high quality of South African IT personnel is recognised by international recruitment organisations such that the poaching of IT personnel is commonplace. Against these trends towards brain drain, there emerges a picture that the 'Department of Home Affairs is a major obstacle to South African enterprises urgently seeking to recruit high level skilled personnel' from overseas (ibid., 2000, p. 39).

The future prosperity of the IT economy in South Africa is clearly linked to the need to address the several issues surrounding the emigration of professional IT skills and to the need to change national immigration attitudes. If successful, there is potential for South Africa to replicate the experience of India, which has benefited considerably from a 'reverse brain gain' as former emigrés are drawn back home to sustain its own developing IT economy (Addison, 2001). Overall, as argued by McDonald and Crush (2000, p. 8): 'In a world of global head-hunting for increasingly mobile skilled personnel, South African policy makers and immigration bureaucrats will need to be much more sophisticated, transparent and strategic in their efforts'. During March 2001 there emerged the first welcome signals that a reversal in the policy paradigm regarding international migration is taking root with an end to the old restrictive immigration policies concerning skilled migrants and steps towards retaining and pro-actively seeking skilled personnel, not least in the IT economy (Kaplan, 2001).

Current Policy Frameworks

The most important output of the SAITIS project linked to the Department of Trade and Industry is the preparation of a detailed policy framework for guiding the development of South Africa's IT economy (SATIS-DTI, 2000). It was acknowledged that in an era of globalisation and increased competition 'the main challenge for countries outside of the developed world is to become increasingly proactive rather than reactive in the development of their indigenous ICT sectors' (SAITIS-DTI, 2000, p. 1). Accordingly, the overarching goal of this framework is to nurture 'a robust, growing and sustainable South African ICT sector that directly supports, and contributes to, the GEAR challenge of sustainable economic growth, social upliftment and empowerment' (SAITIS-DTI, 2000, p. 2). There are two main thrusts to the framework, namely 'the development of the ICT sector and exploiting the capabilities of ICT in developing other sectors of the economy' (SAITIS-DTI, 2000, p. 19).

Concerning the direct upgrading of the IT economy towards the goal of achieving a 'robust, growing and sustainable South African ICT sector with equity', several different objectives and strategies are highlighted. In particular, four major objectives are put forward for developing South Africa's IT economy. First, the need to build local capacity in terms of the key players in the economy, namely large locally-owned enterprises, state-owned enterprises, MNCs and SMMEs. Second, the framework seeks to establish the enabling policy and support environment that would make South Africa an attractive environment for investment in the IT economy. Third, the framework aims to build a world-class sector support infrastructure for the IT economy. Finally, it is aimed to promote exports and facilitate capture of a growing proportion of the global IT market by South African enterprises (SAITIS-DTI, 2000, p. 23).

Of the several proposed strategies and actions that are suggested to fulfil these four objectives the following may be highlighted as perhaps the most significant.

- the SAITIS-DTI study recommends the establishment of a governance structure that will move the proposed strategy into implementation and operate with a clear vision and mandate;

- in terms of generating new growth, alongside broad sectoral development, considerable emphasis is given to strengthening the development of existing IT clusters in Gauteng and Cape Town;
- much attention is to be given to building and/or strengthening linkages among existing players. More especially, a number of proposals are offered regarding empowerment intra-sector linkages, such as supplier development linkages between large companies and SMMEs, outsourcing of non-core functions and mentoring linkages between large enterprises and new entrants;
- a critical strategy in terms of galvanising increased employment opportunities is the 'promotion and support of new entrepreneurial entrants' (SAITIS-DTI, 2000, p. 26). Support actions would include the fostering of incubator/entrepreneur centres at universities and technikons, increased venture capital availability for new start-up enterprises or mentorship support for new entrants;
- creating an enabling environment for investment requires further national government action in a range of areas such as e-commerce, telecommunications policy and not least international migration policy. In addition, government needs to revisit several of its programmes designed to support the manufacturing sector and to refine them in a manner suitable for supporting development of the information economy;
- interventions are encouraged to support the provision of an infrastructure for the IT economy in terms of an enhanced physical infrastructure, greater support for innovation and programmes that support financing for new SMME development. In terms of equity issues the strategic application of government procurement programmes is recommended;
- it is acknowledged that growth of the local IT economy 'cannot be promoted through addressing the local market alone' (SAITIS-DTI, 2000, p. 30) and that export penetration will be important for the future economy health of South Africa's IT economy. A number of proposals are offered pertaining to strategies supporting export expansion, such as creating greater awareness of opportunities and improved marketing and promotion of South Africa's IT strengths;
- finally, two dedicated institutions have been established to assure that South Africa does not lag behind the rest of the world in terms of a digital divide, namely the Presidential National Commission on Information Society and Development and the Presidential Task Force on Information Society and Development. Both of these structures involve public and private sector representatives, including the Chief

Executive Officers of major international corporations in ICT (Dobson, 2001).

Conclusion

The aim in this chapter was to investigate the growth, location, key challenges and policy frameworks that confront South Africa's emerging IT economy, which is both vibrant and growing rapidly. It is evident from the discussion that the national IT economy is currently entering a period of considerable policy flux with a range of new initiatives taking place at both national and regional levels. For economic geographers greatest interest centres upon the changing ownership structures and spatial patterns of IT enterprises and development. At regional level, considerable importance is attached by the national framework to supporting further South Africa's two existing major clusters of IT activity in the Western Cape and Gauteng. In terms of the future economic geography of South Africa, given the dynamism of the IT economy, the aggressive competition played out between these two provinces for IT dominance will be an important factor in shaping the directions of the national space economy.

References

Addison, G. (2000), 'Technology', *Leadership*, September, pp. 20-25.

Bidoli, M. (2000) 'Dimension Data: It's no Ordinary Company', Supplement to *Financial Mail*, 30 June, pp. 6-7.

Bisseker, C. (2001), 'The Barn that Incubates Tech Babies in Cape Town', *Financial Mail*, 2 March.

Bourgouin, F. (2002), 'Information Communication Technology (ICTs) and the Potential for Rural Tourism SMME Development: the case of the Wild Coast', *Development Southern Africa*, vol. 19, in press.

Brown, M., Kaplan, D. and Meyer, J-B. (2000), 'The Brain Drain: an Outline of Skilled Emigration from South Africa', *Africa Insight*, vol. 30, pp. 41-47.

Campbell, K. (2001) 'Telecom 'Duopoly' would put SA at Risk', *Engineering News*, 10-16 August.

Castells, M. (1996), *The Information Age: Economy, Society and Culture: Volume 1 - The Rise of the Network Society*, Basil Blackwell, Oxford.

Castells, M. (1998), *The Information Age: Economy, Society and Culture: Volume 3 - End of Millennium*, Basil Blackwell, Oxford.

City of Cape Town (2001), *Cape Town's Economy: Current Trends and Future Prospects*, City of Cape Town.

Creamer, T. (2001), 'Can SA and Africa Bridge the Digital Divide?', *Engineering News*, 2-8 April.

Crush, J., McDonald, D., Williams, V., Mattes, R., Richmond, W., Rogerson, C.M. and Rogerson, J.M. (2000), *Losing Our Minds: Skills Migration and the South African*

Brain Drain, Cape Town and Kingston, Southern African Migration Project, Migration Policy Series No. 18.

Dobson, W. (2001), 'Driving Growth – from Macroeconomic Stabilisation to Microeconomic Reform', *Sisebenza Sonke*, Issue 1, Article 3.

Dorfling, C. (1999), 'Gauteng Gets Ready to Mine IT Highway', *Engineering News*, 2-8 April.

DTI (1998), 'Special Economic Zones Project: a Spatial Development Initiative in Gauteng', unpublished report, Department of Trade and Industry, Pretoria.

Erwin, A. (2001), '"Activists in a Cause": The DTI revolution', *Sisebenza Sonke*, Issue 1, Article 1.

Franz, L. (2000), ' SA IT Sector Growing at 10% a Year', *Engineering News*, 3-9 March.

Gauteng Province (1997), *Trade and Industry Strategy 1997*, Department of Finance and Economic Affairs, Johannesburg.

Gillwald, A. (2000), 'Building Castells in the Ether', *Link i.t.: The Official Newsletter of the Link Centre*, July, pp. 4-5.

Hodge, J. (2000), 'Liberalising Communication Services in South Africa', *Development Southern Africa*, vol. 17, pp. 373-87.

James, T. ed (1999), 'SAITIS Baseline Studies: A Survey of the IT Industry and Related Jobs and Skills in South Africa', unpublished report prepared for the International Development Research Centre (available at www.saitis.co.za).

James, T. (ed) (2000), 'SAITIS Baseline Studies: Executive Summary: A Survey of the IT Industry and Related Jobs and Skills in South Africa', unpublished report prepared for the International Development Research Centre (available at www.saitis.co.za).

James, T., Esselaar, P. and Miller, J. (2001), 'Towards a Better Understanding of the ICT Sector in South Africa: Problems and Opportunities for Strengthening the Existing Knowledge Base', paper presented at the Annual Forum, Trade and Industrial Policy Strategies, Muldersdrift, 10-12 September.

Kaplan, D. (2001), 'Investment and its Determinants in South Africa – an Economic Policy', *Sisebenza Sonke*, Issue 1 Article 4.

Mabogunje, A. (2000), 'Global Urban Futures: an African Perspective', *Urban Forum*, vol. 11, pp. 165-83.

Mattes, R., Crush, J. and Richmond, W. (2000), ' The Brain Gain and Legal Immigration to Post-apartheid South Africa', *Africa Insight*, vol. 30, pp. 21-30.

Mattes, R. and Richmond, W. (2000), 'The Brain Drain: What do Skilled South Africans Think?', *Africa Insight*, vol. 30, pp. 10-20.

McDonald, D. and Crush, J. (2000), 'Understanding the Scale and Character of Skilled Migration in Southern Africa', *Africa Insight*, vol. 30, pp. 3-9.

Moodley, S., Morris, M. and Kaplinsky, R. (2001), *E-Business: A Necessary Condition for Global Competitiveness*, Policy Brief No. 6, School of Development Studies, University of Natal, Durban.

Ord, J. (2001), 'The Future According to Ord', *Finance Week*, 5 January.

Republic of South Africa (2000), *Green Paper on E-Commerce: "Making it your Business"*, Department of Communications, Pretoria.

Rogerson, C.M. (2000), 'Manufacturing Change in Gauteng, 1989-99: Re-examining the State of South Africa's Economic Heartland', *Urban Forum*, vol. 11, pp. 311-40.

Rogerson, C.M. (2001), 'Knowledge-based or Smart Regions in South Africa', *South African Geographical Journal*, vol. 83, pp. 34-47.

Rogerson, C.M. and Rogerson, J.M. (2000), 'Dealing in Scarce Skills: Employer Responses to the Brain Drain in South Africa', *Africa Insight*, vol. 30, pp. 31-40.

SAITIS- DTI (2000), *South African ICT Sector Development Framework*, SAITIS, Pretoria.

Spicer, D. (2001), 'BEEs Gain Muscle in IT', *Engineering News*, 17-23 August.

Stavrou, A., May, J. and Benjamin, P. (2000) 'E-commerce and Poverty Alleviation in South Africa: an Input to the Government Green Paper', unpublished paper prepared for input to the E-Commerce Green Paper Process.
Stones, L. (2001) 'Didata Set to expand Dutch Operations', *Business Day*, 17 October.

8 Public Transport in the Changing South Africa, 1994-2000

MESHACK KHOSA

> We recall actions of many commuters and transport workers as part of our resistance and struggle against apartheid. We remember these events and struggles not just because they were part of our struggle for freedom, but to remind us that our task today is to build a transport system which is about meeting needs, and not about using transport as an instrument of monopolisation or artificial social engineering (Dullah Omar, Minister of Transport, address to the National Council of Provinces, presenting the National Land Transport Transition Bill, 13 April 2000).

Introduction

Popular transport struggles in South Africa were at the heart of the struggle for liberation (Khosa, 1995), as is clear from the citation of Dullah Omar above. Hence the passage of the National Land Transport Transition Act in 2000, described by Omar (2000a) as 'a significant milestone on the road to the fundamental transformation of transport in our country', marked an important chapter in the history of the South African transport sector.

After more than six years of consultations and negotiations, the National Land Transport Transition Act was the first important transport legislation introduced by the ANC-led government. However, the implementation of this Act poses significant obstacles. Changing the nature of land-use planning, road space management, planning and regulation, and subsidy targeting will need agreement on the objectives as well as the political will to follow through on the objectives. In addition, the co-ordination of the relevant departments at all levels of government with a view to applying the Act is a complex task.

The legal framework for the establishment of transport authorities has now been set. A general principle in terms of the Act is that land transport planning, which will be the responsibility of transport authorities, must be integrated with land development. The National Land Transport Transition Act prescribes that transport plans form an essential part of integrated development plans (Omar, 2000b). However, the Act raises more questions than it answers. This chapter therefore analyses the changing nature of transport policy since 1994 together with public perceptions of public transport, tarred roads and street drainage. Two Human Sciences Research Council surveys (HSRC, 2000a and 2000b) are compared to determine public perceptions of transport.

Before the first democratic election in 1994, the African National Congress (ANC) developed its radical Reconstruction and Development Programme (RDP) as an election manifesto, and later proclaimed this programme an instrument of fundamental change in the new South Africa. However, less than two years after the new government assumed power, a neo-liberal economic policy was introduced, which in essence replaced the initial programme (Bond, 2000). The nature of the new policy caused heated debates, with the labour movement and some civil society organisations accusing the government of 'selling out the people's mandate' to the private sector and the International Monetary Fund. Nevertheless, the ANC-led government published several statistics after 1995, arguing that its social delivery programme, especially service delivery, was successful, and that the goals of the RDP were still central to government policy (Bond, 2000).

Background and Argument

Post-apartheid South Africa has inherited a public transport system that is under-performing when compared with the national reconstruction goals as outlined in various government policy documents. RDP goals including basic mobility, basic access and social integration, which various government officials still claim are guiding their development programmes, are generally not met. Workforce mobility is still restricted, creating friction around national efforts to create employment opportunities. The current spatial distribution of the transport system still leaves commuters and residents without basic services, and the overall inefficiency of the system induces the provision of transport subsidies, as was the case in the past. Several studies have concluded that few households in South Africa have access to a private car, while more than 60 per cent of the 'ultra poor' walk to work (Khosa, 1998). Transport services for the poor are limited and inappropriately located, and poor people often have to use more than one mode of transport to reach their destination.

This chapter outlines the evolution of transport policy since 1994, reviews achievements in implementing the RDP, and evaluates public perceptions of the delivery of public transport and tarred roads and drainage. It is argued that, instead of the poor and rural people, middle- and high-income earners and urban and metropolitan residents are the major beneficiaries of good quality services in respect of public transport, tarred roads and street drainage.

Transport Policy Formulation since 1994

Policy-making has changed dramatically since 1994. Whereas past transport policies were imposed on a disenfranchised majority, the new government has embarked upon consultation with civil society as a participant in policy formulation. In addition, urban transport policy in South Africa is shaped by dynamic global processes such as deregulation, commercialisation and privatisation on the one hand, and contradictory local features such as remnants of apartheid administration and the imperative for inclusive and participatory decision-making on the other hand. Despite the obstacles imposed by these forces, various landmarks have been achieved in the transformation of public transport in South Africa which are considered in this chapter.

The National Transport Policy Forum

The National Transport Policy Forum (NTPF) was launched in February 1992 to bring together a range of interest groups, many of whom had been excluded from the formulation of transport policy in the past. The NTPF included the ANC, the South African National Civic Organisation (SANCO), the Congress of South African Trade Unions (COSATU), the Congress of Traditional Leaders of South Africa (CONTRALESA), the National African Federated Transport Organisation (NAFTO), the Pan-Africanist Congress (PAC), the Southern African Black Taxi Association (SABTA) (the first national African taxi organisation) and TRANSNET (a state-controlled company representing the South African Airways and the South African Railways and Harbours, established in 1990), and several organisations representing the private sector.

Through the NTPF the formulation of transport policy was publicly discussed by 'credible' representatives. In fact, the NTPF described access to transport as a basic human right. After two years of discussion, a 'people-centred' transport policy was published by the NTPF in September 1994 (National Transport Policy Forum, 1994). This policy document outlined strategies to overcome fragmentation within the transport sector (ibid., pp. 13-14). In addition, transport was recognised by the NTPF as an instrument of social transformation: 'The transport industry should be used as an instrument of transformation. Emphasis should therefore be placed on the creation of new businesses and empowerment as a tool in the economic process' (ibid., p. 2). The co-ordinated development of transport policy by the NTPF was a radical deviation from the development of previous transport policy, as it was a result of debates, consultations and consensus among various stakeholders.

Transport and the ANC's Election Manifesto

Although the ANC was a member of the NTPF, its involvement in the working groups and deliberations was insignificant. Indeed, the ANC was even absent from the launching of the National Transport Policy Forum's policy document in September 1994. The liberation movement in general and the ANC in particular had not prepared a single comprehensive document on transport policy when the first general election took place in 1994, in contrast to the health and education sectors. In the well-publicised RDP document, published four months before the election in April 1994, transport occupied only three and half pages of the 147-page document. In the RDP document, the ANC suggested that the new government develop 'an effective publicly owned passenger transport system ... integrating road, rail and air transportation' (African National Congress, 1994, p. 36). In addition, the ANC argued that urban commuters should be encouraged to use public transport, and should be actively discouraged from using cars (by means of parking, access and fuel levies). The ANC declared that access to transport, as to health, education and housing, was a basic human right largely to be met by the new government: 'The needs of women, children, and disabled people for affordable and safe transport are important. Adequate public transport at off-peak hours, and security measures on late-night and isolated routes, must be provided. Additional subsidies for scholars, pensioners, and others with limited incomes will be considered' (ibid., p. 38). However, the evidence suggests that the transport section in the RDP document was based on rhetoric rather than a rigorous analysis of the transport sector in South Africa.

The transformation of public transport policy started with the transport review process, the formulation of the Green and White Papers, and the establishment of the National Taxi Task Team. Each of these steps has contributed to changing public transport policy since 1994.

Transport Policy Review

After the inauguration of Nelson Mandela as first president of the democratic South Africa in 1994, the transformation of the transport sector and the development of transport policy were spearheaded by the 'new' Department of Transport (Department of Transport, 1995). Early in 1995, under the director-general, Ketso Ghordan, and the then Minister of Transport, Mac Maharaj, the Department of Transport embarked on a fundamental revision of transport policy. A steering committee consisting of thirteen representatives from the public and private sectors was established by the Department of Transport to guide the policy review. Six working groups were established to analyse policy issues within specific transport sectors, respectively civil aviation, infrastructure, land freight transport, land passenger transport, road traffic management and shipping. The reports of these sectoral working groups were submitted to the first transport plenary meeting in July 1995, attended by 300 representatives from private and public institutions and other transport agencies.

In October 1995 the land transport working group published its draft passenger transport policy framework. It stated that passenger transport was insufficiently provided for by legislation, and that the little passenger transport legislation that existed was administered by various authorities at different levels of government. The land transport working group suggested that a single, national land transport act be drafted to clarify the roles and functions of the various levels of government.

Following the first transport policy plenary meeting and several other seminars held by the sectoral working groups, a Green Paper on national transport policy was drafted. This was further discussed at the second transport policy plenary attended by some 300 public, private and civil society sector representatives in February 1996. The Green Paper was published in March 1996 (South Africa, 1996a). Further inputs and submissions were incorporated in April and May 1996. The *White Paper on National Transport Policy* was finally completed and accepted by cabinet in September 1996 (South Africa, 1996b).

White Paper on National Transport Policy, 1996

According to the White Paper the transport policy aimed to '[p]rovide safe, reliable, effective, efficient, and fully integrated transport operations and infrastructure which will best meet the needs of freight and passenger customers at improving levels of service and cost in a fashion which supports Government strategies for economic and social development whilst being environmentally and economically sustainable' (South Africa,

1996b, p. 3). In the past, government's dominant role was seen as that of a 'regulator of bureaucratic detail[s], a provider of infrastructure, and a transport operator, but has been weak in policy formulation and in strategic planning' (South Africa, 1996b, p. 7). The government intended to reverse this. The means for doing so would be to focus on policy and strategy formulation as its prime role and regulation as its prime responsibility, and to reduce its direct involvement in operations and the provision of infrastructure and services to allow for a 'more competitive environment' (South Africa, 1996b, p. 7).

The new constitution and the White Paper introduced radical changes in transport. Schedule 4 of the Constitution of the Republic of South Africa allocated concurrent responsibility for public transport, road traffic regulation and municipal public transport to the national and provincial governments. Transitional metropolitan and local councils would be responsible for metropolitan co-ordination, land usage and transport planning; arterial metropolitan roads and storm water drainage; public transport services; and traffic matters (Department of Transport, 1995). Each province was empowered to legislate on public transport, within a broad national policy framework, and to determine its own transport policies, which should also guide metropolitan and local transport authorities. Although the constitution allowed for concurrent powers, the transfer of bus services, ports, airports and railways to the provinces has been delayed to enable the provinces to establish the necessary capacity and make institutional arrangements.

The White Paper promised to offer financial and technical assistance to minibus taxis in order to 'improve their financial viability' (South Africa, 1996b, p. 24). With regard to rail passenger transport, the White Paper promised to 'end the deficit financing system and replace it with a concession system which will ensure more efficient and effective use of funds' (South Africa, 1996b, p. 25). It also recommended 'regulated competition' in bus operations. Since the 1940s the bus industry had been monopolised by a few companies. For example, PUTCO was the recipient of up to 45 per cent of the annual R815m. bus subsidies, with some 35 other bus companies sharing the rest (Department of Transport, 1996a). In 1996/97 bus subsidies from the government and local authorities amounted to R967m., subsidising 20m. passenger trips (SAIRR, 1998). The White Paper committed the government to ending the subsidy system and replacing it with interim contracts and tenders (South Africa, 1996b, p. 24).[1] The White Paper also suggested that the right to public transport should become the cornerstone in the transport policy of future governments and that transport planning would thus require a totally new approach.

The central tenet of the White Paper was that all freight and passenger transport operations should be run on a commercial basis rather than as a social service (South Africa, 1996b). This thrust marked a complete reversal of the ANC's policy position outlined in the RDP document. This change in policy was also reflected in other national policies, particularly the macro-economic strategy (GEAR), which was released in June 1996. One dramatic change occurred in the Department of Transport (DoT) itself. The number of people employed in the Department was reduced from 1,400 to about 250 (Maharaj, 1999). Specialist divisions were created at arm's length, namely the aviation authority, the maritime authority, the road authority and cross-border agencies. The DoT is now focused on transport policy development, strategic planning and implementation, and regulation and safety.

National Taxi Task Team

Two other transport policy initiatives ran parallel with the process that led to the drafting of the White Paper, namely the National Taxi Task Team (NTTT, established in March 1995 and consisting of nine taxi industry representatives, nine government representatives and nine specialist advisors) and the Task Team on the Restructuring of State Owned Enterprises.

The objectives of the NTTT were to investigate problems facing the taxi industry and formulate solutions and/or options to ensure the short and long-term sustainability of the industry (South Africa, 1996). Between August and December 1995 the NTTT conducted public hearings in all the provinces to obtain input on problems in the industry. Over 90 days of public hearings on the taxi industry took place, and some 32 venues were used. The NTTT submitted a report to the Department of Transport (NTTT, 1996), and cabinet approved the recommendations the same year. These included regulation and control of the taxi industry, restructuring the industry into more formal business units or co-operatives, and economic assistance through a 'short-term survival package'. Based on these recommendations, each of the nine provinces established a provincial taxi office and appointed a taxi registrar. The government committed itself to providing financial assistance to the taxi industry in various forms to facilitate the establishment of taxi co-operatives and training in business skills. Some of the key NTTT recommendations were incorporated into the *White Paper on National Transport Policy* (South Africa, 1996b). Provincial taxi offices and provincial registrar offices were established in 1997 and 1998.

Minibus taxis transport over 65 per cent of public transport users in South Africa. There are 80,000 operators with 127,000 vehicles with a turnover of R11m. per day. One of the recommendations of the Moving South Africa (MSA) project is a recapitalisation programme, aimed at transforming the taxi industry over a period of five years. In essence this programme entails the replacement of the current 15-seater taxis with 18- and 35-seater taxis specifically designed to comply with the high safety standards for public passenger transport. Over the next four years the government could provide up to R4bn. in the form of scrapping allowances of up to 30 per cent of the cost of a new vehicle. However, in terms of the National Land Transport Transition Act, the government will be able to prescribe the use of the new vehicles only in 2003 (*Pretoria News*, 6 April 2001).

A task team representing the Department of Transport, the Department of Trade and Industry, the Department of Finance and the Department of Minerals and Energy developed the recapitalisation programme. The tender for 18- and 35-seater vehicles was issued in 1999 and Iveco and AMC from South Africa, DaimlerChrysler from Germany, Gaz from Russia, Tata from India and Kwoonchung from China were shortlisted (*Pretoria News*, 6 April 2001). The South African Taxi Council (SATACO), the 'parliament of the taxi industry', now officially represents the taxi industry in the recapitalisation programme (Omar, 2000a).[2]

Moving South Africa – the Legacy of Maharaj

The DoT ran the Moving South Africa (MSA) project from June 1997 to October 1998 (Department of Transport, 1998b and 1999). This was the first government-initiated study in the history of transport in South Africa that considered all transport modes including rail, road and air transport. The project sought to develop a 20-year strategy to achieve the goals of the White Paper. The ultimate aim was to ensure that the transportation system of South Africa meets the needs of the 21st century and therefore contributes to the country's growth and economic development. The DoT charged MSA with responsibility for helping to break new ground in government approaches to long-term strategic issues. Hence MSA undertook to identify and clarify the nexus between policy and strategy (Maharaj, 1999). Since the White Paper had already put forth the vision, MSA's mission was to determine how to implement that vision in a way that would be consistent with the key thrusts articulated above, in an environment of limited resources, capacity and time. The strategy had to be based on the objectives of the White Paper, and where necessary to reconcile or choose between competing objectives articulated in the White

Paper. In addition, the strategy had to create a context for action towards achieving the White Paper objectives.

The vision of the MSA is bold, but what distinguishes it from the Reconstruction and Development Programme (RDP) and the *White Paper on National Transport Policy* is its shameless use of neo-liberal language and its commitment to user charges. Scattered throughout the MSA report passengers are referred to as 'customers' and 'clients', which clearly indicates a market-orientated approach to public passenger transport in South Africa.

In order to assess progress in the transformation of the transport sector, the following section outlines the first RDP policy audit.

Transport Policy Audit, 1994-1999

One of the most important contributions to the assessment of transformation in South Africa was the commissioning of the RDP policy audit. The RDP policy audit involved assessment of infrastructure and service delivery based on official documents and in-depth interviews. Official claims in terms of infrastructure and service delivery were assessed on the basis of the RDP 'mandates' (Bond and Khosa, 1999; Khosa, 2000b). This section of the chapter seeks to use aspects of the RDP policy audit on public transport and a series of HSRC surveys on public perceptions on the delivery of public transport in South Africa to assess the transformation in this sector.

Expansion of Public Transport

The RDP emphasised the need to expand affordable public transport options, especially rail. This is captured in various sections of the RDP policy document of 1999:

> 2.9.3 An effective publicly owned passenger transport system must be developed, integrating road, rail and air transportation ... A future transport policy must: promote coordinated, safe, affordable, public transport as a public service ...
> 2.9.5 As a first priority, rail transport must be extended (African National Congress, 1994).

The Department of Transport's early policy papers - such as *Working Documents for Land Transport Bills and Cross-Border Road Transport Bill* - reassessed state ownership of transport services and advocated corporatisation of municipal services, including rail. The *Working*

Documents policy targets for transport access were ambitious, and more detailed than those offered in the RDP. Recent policies effectively support privately owned transport and restrict the state to planning, regulation and control.

Regulation of Private Transport

The RDP aimed to counter individual car use by imposing higher taxes and allocating larger public transport subsidies:

> 2.9.5 Commuters should be encouraged to use public transport, and should be actively discouraged from using cars (via parking, access and fuel levies). The funds so raised must be used to directly benefit the provision of public transport ...
> 2.9.11 Funding for public transport would come both from central government and from local rates and taxes. The (Metropolitan Transport Authorities) must be empowered to impose such levies and taxes as may be appropriate and the funds thus raised must be used primarily to promote public transport (African National Congress, 1994).

Taxes and constraints to car use were endorsed in the 1995 *Urban Development Strategy*, but subsidies were discouraged and a user-pays principle recommended.

> Transportation subsidies alone cost the fiscus in excess of R2bn. a year, and are particularly excessive in providing for the inhabitants of far-flung 'commuter townships'... As far as possible, infrastructure should be funded through user charges and/or investments by the private sector ... Unrestrained car usage and subsidised car parking should be contained through the application of policy instruments including strict parking policies, access restrictions for private cars, higher licence fees, road pricing and area licensing ... The application of funds to transport improvements should be self-sustaining and replicable. To encourage this, the users of urban transport facilities should pay for all or most of the costs incurred within the limits of affordability (South Africa, 1995, pp. 15, 18-19, 24, 28).

Formal transport policy was silent on the means of effecting an incentive shift. The RDP also mandated stronger regulation of private transport: 'All privately-controlled passenger transport must be effectively regulated and controlled' (ANC, 1994, section 2.9.3). However, the policy favoured self-regulation. According to the National Land Transport Transition Bill:

The roles and responsibilities of the key stakeholders and service providers will be clearly agreed. This will enable government regulation to be kept to a minimum, while the private sector will be able to build and operate within a competitive environment, be socially and environmentally responsible and self-regulating (Department of Transport, 1998c, p. 6).

This new trend excluded taxi transport, which was increasingly subjected to government intervention. The National Road Traffic Act (No. 93, 1996) tightened other transport laws.

Transport Planning

The RDP stressed decentralised control of transport planning: 'A future transport policy must: ensure accountability so that the people have control over what is provided' (ANC, 1994, section 2.9.3). At the same time it called for more comprehensive, integrated planning: 'A future transport policy must: ensure comprehensive land-use/transport planning' (ibid.). Accountability was affirmed in the 1996 White Paper, as was integrated transport planning in the 1998 document *Moving South Africa*. The RDP proposed a single transport agency to co-ordinate planning and financing:

> 2.9.9 For all public transport services to be fully integrated their functioning must be coordinated and financed by one organisation. The organisation should be accountable to the public and responsible for the provision, coordination and funding of all public transport and the infrastructure necessary for public transport (in cooperation with the national public works programme). The organisation should specifically address current problems such as uncoordinated tariff structures, duplication of services, and conflict as a result of different forms of ownership. Minimum norms and standards, policy frameworks and the format of transport plans for national, provincial, urban and rural areas should form an integral part of the responsibilities of this organization (ANC, 1994).

While subsidies to bus and rail transport remained substantial (R2bn. p.a.), and while registered use of public rail transport increased by slightly more than the growth in population, the aggregate amounts available declined. Indeed, the transport budget was several billion rand per year short in terms of capital spending (particularly for new roads) due to fiscal constraints. The responsibility to generate additional resources is now placed on public-private partnerships, with the national road transport agency charged with the responsibility to manage, renovate and maintain the national road network.

Although the National Land Transport Transition Act, enacted in 2000, provides a framework for integrated co-ordinated planning, there is recognition that this may take at least five years, as the transformation of local government, which includes local government's taking responsibility for transport, still has to take its course. Intersectoral collaboration in government, policy formulation and programme implementation still leaves much to be desired. There is apparently little commitment to speeding up delivery through 'sustainable action'.

Special Transport Needs

The RDP mandated additional transport support for people with special needs:

> 2.9.3 A future transport policy must take into account the transport needs of disabled people ...
> 2.9.13 The needs of women and children for affordable and safe transport are important. Adequate public transport at off-peak hours, and security measures on late-night and isolated routes, must be provided. Additional subsidies for scholars, pensioners and others with limited incomes will be considered (ANC, 1994).

Changes in transport policy for disabled people, women, children, scholars and pensioners have not been specified, but the RDP placed special emphasis on rural transport.

> 2.9.2 Rural areas require more frequent public transport and improved facilities, at an affordable cost. There is inadequate access for emergency services in rural areas, inadequate public transport frequencies and route coverage, poor coordination, and other inefficiencies. Indeed, in many rural areas there is no public transport at all (ANC, 1994).

The commitment to rural areas was repeated in other policy documents including like the 1996 *Green Paper on National Transport Policy* (Department of Transport, 1996a) and the *Rural Development Strategy* (South Africa, 1995b).

Deviation from the RDP

Evidence in the preceding section suggests that the RDP proposals were fundamentally revised in the first term of office of the ANC-led government. Transport is no longer seen as an essential service, but as a commodity in the market place. In the words of Dullah Omar, Minister of Transport:

We must strive to create a competitive and market-based transport environment. This will facilitate and promote greater participation by the private sector in the delivery of transport services, and generate the appropriate market forces to regulate supply and demand pressure. This will affect modal choices as well as the development of truly market-driven inter-modal transport (Omar, 2000a).

These shifts are also evident in sectors other than transport. The impact of these changes has not yet been thoroughly analysed. An analysis that uses a variety of methods will assist in teasing out some of the key aspects of service delivery in general, and transport delivery in particular.

Perceptions of Changes in Service Delivery at Local Level

Since 1994 policy analysts have used various yardsticks to measure progress in terms of service delivery (Khosa, 2000a). For example, the *Mail & Guardian* provides a (largely superficial) report card for each cabinet minister, and *RDP Monitor* produces a monthly report on the implementation of the RDP. The discussion below takes as its starting point the results of HSRC surveys carried out in November 1999 and September 2000.[3]

The first democratic government elected into power in 1994 inherited high levels of unemployment, huge income inequalities and astronomical service backlogs. One of the key programmes introduced by the government was the RDP, which sought to address some of the national problems. By doing so, the government expected to kick-start economic growth, which was negative at the time. Critical aspects identified for improvement were running water, affordable housing, electricity, health and education. However, in less than two years, the ANC-led government abandoned this programme, and replaced it with a neo-liberal development programme. This section of the chapter seeks critically to appraise the extent to which the public perceived government-sponsored development to be effective, especially in November 1999 and September 2000.

Perceptions of Public Transport

The legacy of apartheid is still reflected in the transport sector. The location of the majority of Africans, coloureds and Asians further away from their workplaces than whites resulted in longer travelling distances for the former. In addition, impoverished areas had few access and exit routes. Transport spending was inadequate and ill targeted. For instance, the beneficiaries of transport subsidies were not the poor, but middle-income earners. Also, rail transport was the most heavily subsidised mode of transport, but carried fewer commuters than other modes of transport. People in rural areas were the most disadvantaged in terms of access to roads and transportation (Khosa, 2000a).

Since 1994 there has been a shift from spending on defence and transport to spending on social services. Poor access to transport impacts daily on the quality of life of most people. Inordinate travelling times and distances also constrain an individual's ability to participate in family and community life and leisure activities, as well as society's involvement in economic processes. Workers are often dependent on inefficient transport to reach the premises of their employers, and consumers often have difficulty in getting to retail stores. In other words, if transport is non-affordable and inefficient, people's participation in production and consumption cannot be maximised. This, in turn, will prevent them from raising their quality of life, as participation leads to job creation, economic development and growth.

Between November 1999 and September 2000 the proportion of respondents indicating an improvement in public transport decreased from 39 per cent to 32 per cent, and the proportion of respondents indicating a deterioration increased from 25 per cent to 31 per cent. The increased proportion of people who perceived the public transport service to be deteriorating implies that transport delivery has been declining in the post-apartheid period. Although public transport was the most visible area of popular struggle against the apartheid regime, little has been invested in it by way of finances and concerted effort since 1994.

Perceptions of Public Transport by Race

Whereas Africans and whites reported public transport service deterioration, coloureds and Asians reported public transport improvement. White perceptions of deterioration can be attributed to their previous public transport privileges, especially when municipalities still provided basic transport services. Whereas negativity among Africans and whites increased (from 18 per cent to 31 per cent among Africans, and from 30 per cent to 41 per cent among whites), negativity declined among coloureds (from 30 per cent to 23 per cent) in the same period. Africans were however the major users of public transport, with the majority of whites owning private cars (Table 8.1).

This overall pattern implies that the majority of users of transport - mainly Africans and mainly the poor - perceived a reduction in the level of improvement between November 1999 and September 2000.

Table 8.1 Perception of Change in Public Transport Delivery, 1994-2000 (percentage)

Perception	African		Coloured		Asian		White	
	Nov. 1999	Sept. 2000	Nov. 1999	Sept. 2000	Nov. 1999	Sept. 2000	Nov. 1999	Sept. 2000
Improved	48	37	23	26	23	42	23	11
Same	32	31	42	48	42	33	42	34
Worsened	18	31	30	23	30	21	30	41
Uncertain	2	1	5	3	5	4	5	14
Total	100	100	100	100	100	100	100	100

Perceptions of Public Transport by Area

Public transport improvement appears to have been more evident in urban and metropolitan areas than in rural areas. Some 40 per cent of rural respondents discerned a worsening in public transport services compared with 25 per cent who felt there was an improvement. This suggests that public transport improvement largely occurred in urban and metropolitan areas (Table 8.2).

Table 8.2 Perception of Improvement in Public Transport by Area (percentage)

Rating	Metropolitan	Urban	Rural	SA average
Improved substantially	5	6	2	4
Improved	25	35	23	28
Same	30	36	32	33
Worsened	21	15	27	21
Worsened substantially	13	6	13	10
Uncertain	5	4	2	3

Poverty was much more evident in rural areas, with the majority of residents forced to walk long distances to get access to public transport. This was a major disadvantage for rural people as public transport was central to accessing basic services and commercial outlets. Urban and metropolitan residents had little difficulty in this regard.

Perceptions of Public Transport by Province

Public perceptions of the delivery of public transport also differed by province. Public transport appears to have improved in Mpumalanga, the Free State and North West between November 1999 and September 2000 (Table 8.3). The greatest proportions of respondents who perceived a deterioration in public transport were located in Gauteng, the Northern Province, KwaZulu-Natal and the Eastern Cape. As the latter three are the poorest provinces in South Africa, the finding confirms that service provision did not benefit the poor.

Table 8.3 Perception of Improvement in Public Transport by Province (percentage)

Province	Improved	Same	Worsened	Uncertain
Western Cape	27	47	20	7
Eastern Cape	31	28	37	4
Northern Cape	26	47	21	6
Free State	35	30	33	3
KwaZulu-Natal	30	33	37	1
North West	37	42	16	5
Gauteng	32	26	38	4
Mpumalanga	55	21	23	2
Northern Province	28	34	36	2
SA average	32	32	31	3

Slightly more than a third of Gauteng respondents perceived a deterioration in public transport, compared with a third who perceived an improvement. The Gauteng respondents who perceived a deterioration in public transport were largely resident in African townships, informal settlements and peri-urban areas, rather than in the historically 'white' suburbs where the majority of people used private transport.

Perception of Public Transport Service Delivery by Living Standard

Perceptions of the provision of public transport also differed by level of wealth. Those who could afford to pay the fares largely used public transport (it was also far cheaper to use public transport than a private car). Evidence from the September 2000 survey suggests that those who believed that public transport had improved fell between Living Standard Measure (LSM) 4 and LSM 6, with the rest indicating a deterioration. The September 2000 survey suggests that the beneficiaries of better service delivery were not the poor, but middle- and high-income earners who were also able to pay for the services. The findings challenge the mainstream perspective, which suggests that government's programmes are largely benefiting the poor and the marginalised.

Gender and Public Transport

Perceptions of public transport delivery also differed by gender. Women constituted a greater proportion of those who perceived a deterioration in public transport services. The needs and interests of women were not always considered. For example, women traders in urban and rural areas often travelled with heavy and bulky baggage, which ordinary modes of transport did not accommodate. Lack of rural public transport often increased costs for women indirectly.

Although the special needs of women are prioritised in policy and legislation public transport in South Africa is still a dangerous sector for women, especially at dawn, dusk and at night. Special programmes to provide safe and secure transport for women are still a pipe dream. Given the dominance of the neo-liberal perspective, only an alternative development agenda might sway government's approach, though the prospects of this appear dim. Men continue to be the majority of transport operators and owners of transport, even though the majority of transport users are women. Efforts at encouraging emerging transport contractors have not been complemented by efforts to ensure that women are better represented in public transport provision in South Africa.

Tarred Roads and Street Drainage

Lack of funds impacted on the implementation of road policy in South Africa. In September 1999 the Department of Transport indicated that there was a R43bn. backlog in terms of road maintenance, repair and upgrading. Most of South Africa's road network exceeded its twenty-year lifespan, which could result in hazardous road conditions. Eighty-five per cent of roads to rural villages were inadequate, compared with 32 per cent of roads to farming communities. Significant numbers of respondents believed that tarred roads and street drainage had deteriorated. During the September 2000 survey, only 29 per cent of South Africans indicated an improvement in the provision of tarred roads and street drainage in their areas, the same as in November 1999, compared with 41 per cent who discerned a deterioration, an increase from 32 per cent in the same period.

Perceptions of tarred roads and street drainage also differed by race. A higher percentage of whites and Africans than of Coloureds and Asians discerned a deterioration in tarred roads and street drainage. However, the proportion of whites who noted an improvement increased from 8 per cent to 18 per cent between November 1999 and September 2000, and the proportion indicating a deterioration decreased from 44 per cent to 40 per cent (Table 8.4).

Table 8.4   Perception of Improvement in Tarred Roads and Street Drainage, 1994-2000 (percentage by race)

Perception	African		Coloured		Asian		White	
	Nov. 1999	Sept. 2000	Nov. 1999	Sept. 2000	Nov. 1999	Sept. 2000	Nov. 1999	Sept. 2000
Improved	33	24	33	25	27	54	8	18
Same	34	29	43	48	61	30	46	41
Worsened	31	44	20	25	10	13	44	40
Uncertain	2	3	4	2	2	2	2	2

There was a noticeable drop from November 1999 to September 2000 in the proportion of Africans and coloureds who felt that tarred roads and street drainage had improved. The results suggest widespread dissatisfaction with the government's service delivery programmes among the poor and previously disenfranchised (Khosa, 2001).

Public perceptions of tarred roads and street drainage also differed by gender. More men than women indicated that tarred roads and street drainage had improved, and more women than men indicated a deterioration in these services. Not surprisingly, the majority of road contractors and road construction workers were men. Moreover, the majority of access roads built in post-1994 under the community-based public works programmes were destroyed in the floods of 1999 and 2000. Several bridges were washed away in the rural Eastern Cape and Northern Province. Proportionately more women were affected by this, as the destruction of rural road networks and bridges made it difficult for them to fetch water and wood. Access to basic services was also severely affected.

Perceptions of an improvement in tarred roads and drainage also differed by province (Table 8.5). The provinces with the greatest proportion of respondents indicating service deterioration were the Northern Province, North West, the Free State and the Northern Cape. These provinces also had the smallest proportions of respondents indicating service improvements. Interestingly, Mpumalanga, a reportedly scandal-ridden province with an ineffective government, had high proportions of respondents who indicated an improvement in public transport, tarred roads and street drainage.

Table 8.5 Perception of Improvement in Tarred Roads and Street Drainage (percentage by province)

Province	Improved	Same	Worsened	Uncertain
Western Cape	34	49	16	1
Eastern Cape	21	30	49	0
Northern Cape	3	33	45	3
Free State	19	26	51	3
KwaZulu-Natal	30	57	13	1
North West	17	32	44	7
Gauteng	27	35	22	6
Mpumalanga	32	37	28	3
N. Province	10	21	66	3
SA average	25	32	41	3

The greater proportion of respondents in the Eastern Cape and Northern Province indicating service deterioration confirms the perceived continued marginalisation of these provinces, even though official government policy targets poverty-stricken areas for improved service delivery. In KwaZulu-Natal the perceived improvement in the provision of tarred roads and street drainage is to some extent attributable to the high-profile road projects championed by the Minister of Transport in KwaZulu-Natal, S'bu Ndebele. At the national level, the proportion of respondents indicating deterioration in services was greater than the proportion indicating an improvement. As tarred roads and street drainage are provided by provincial and local government, the findings in this chapter alert one to the fact that local and provincial government is not performing optimally. Carefully crafted policy documents at national level remain rhetoric, bearing little resemblance to service delivery at local and provincial level.

Of the overall 41 per cent of respondents who indicated that tarred roads and street drainage had worsened, the majority were rural residents. Dissatisfaction with service delivery in rural areas was largely attributable to the failure of the service delivery machinery to prioritise rural areas. No integrated development rural strategy has been implemented in the seven years after the democratic government was elected. Although initially mentioned in the RDP, and a draft framework for it was circulated in 1995, the *Rural Development Strategy* is still debated. Hence rural households continue to be marginalised and do not share in the social and material benefits of the new dispensation.

Perceptions of tarred roads and street drainage also differed by levels of wealth. Of those who indicated significant service improvements, the majority were respondents falling in LSM 5 and LSM 6, representing the bulk of the middle-class. These results suggest that the lower the LSM the less the improvement in tarred roads and street drainage. Lower LSMs tended to have proportionately more respondents who indicated a deterioration in tarred roads and street drainage. These findings emphasise that low-income households were often overlooked by new national service and developmental programmes, whereas the middle- and high-income households were the major beneficiaries.

Conclusion

The South African passenger transport system was designed largely to transport people between residential townships and their workplaces. Because of the spatial distances between the two, the working poor spent much time and money on transport. So, when transport policy changed in the new political dispensation from exclusion to inclusion, from being drawn by a few technocrats to consulting with civil society in the process of policy formulation, an improvement was expected in the situation of the poor. However, as is evident from this chapter, the proposals in the RDP were fundamentally changed in the first term of office of the ANC-led government, and affected transport as well as the other sectors.

To improve our understanding of the impact of this shift, the results of the HSRC's surveys of November 1999 and September 2000 were analysed in respect of public perceptions of public transport, tarred roads and street drainage. From this analysis, the following observations could be drawn:

- Positive ratings of public transport, tarred roads and street drainage declined between November 1999 and September 2000.

- Rural areas are affected worst by the decline in public transport, tarred roads and street drainage. This was corroborated by the admission of S'bu Ndebele, Minister of Transport in KwaZulu-Natal, during a budget speech in March 2001, that 60 per cent of roads in KwaZulu-Natal were considered to be in a very poor condition (Ndebele, 2000).
- Most of the improvement in public transport, tarred roads and street drainage services appears to have occurred in metropolitan and urban areas, rather than in rural areas where the majority of the poor live. This is based on the finding that most of the respondents who were dissatisfied are from rural areas, and most of the respondents who were satisfied hailed from metropolitan and urban areas.
- Levels of wealth appear to have been a critical aspect in service delivery. The poor do not appear to be the major beneficiaries of improved service delivery. In fact, middle- and high-income earners emerged as the most satisfied and happiest recipients of service delivery. This finding calls for a review of the current delivery mechanisms to ensure that the poor are the primary beneficiaries.

The assessment of transport delivery in this chapter suggests that the current neo-liberal framework for service and infrastructure delivery should be fundamentally reviewed to effect a significant improvement in the quality of life of the poor. Moreover, the lack of political will to ensure efficient and effective transport for the poor bodes ill for the country as a whole. Indeed, the trade union movement has been radicalised and organs of civil society have shown open defiance in the face of the poor quality services provided as well as the government's newly found zeal to promote outsourcing, privatisation and deregulation of economic activities.

Notes

1. In terms of bus contracts, the replacement of lifelong bus operator permits and subsidies has now been completed, and converted from interim to tendered contracts.
2. SATACO was born out of a memorandum of understanding signed by government and the taxi industry in 1999 to set up a unified representative structure for the taxi industry. A splinter group called the National Taxi Alliance, which was initially opposed to taxi recapitalisation, is currently holding unity talks with SATACO. Democratic elections within the taxi industry were to precede SATACO's involvement in taxi recapitalisation.
3. The survey instrument was a formal questionnaire. The questions were arranged around themes and it took respondents between 60 and 90 minutes to complete the questionnaire. A national sample of 2,704 respondents was selected (the realised sample totalling 2,666), grouped in clusters of eight and drawn from 338 census enumerator areas (EAs) as delineated for the 1996 census. The sample by province and lifestyle ensured adequate representation: the HSRC categorised each EA in terms

of one of 23 lifestyles, based on an analysis of the 1996 census data. Disproportionately large samples were selected from areas known to be inhabited by the two smallest population groups, namely areas with predominantly Asian (Indian) or coloured populations.

References

African National Congress (1994), *Reconstruction and Development Programme*, Umanyano Publications, Johannesburg.

Bond, P. (2000), 'Infrastructure Delivery: Class Apartheid', *Indicator SA*, vol. 17 (3), pp. 18-21.

Bond, P. and Khosa, M.M. (1999), *An RDP Policy Audit*, HSRC Publishers, Pretoria.

Department of Transport (1995), *Annual Report of the National Department of Transport, 1995/96*, Government Printer, Pretoria.

Department of Transport (1996a), *Green Paper on National Transport Policy*, Government Printer, Pretoria.

Department of Transport (1996b), *White Paper on National Transport Policy*, Government Printer, Pretoria.

Department of Transport (1998a), *Annual Report of the National Department of Transport, 1997/98*, Government Printer, Pretoria.

Department of Transport (1998b), *Moving South Africa*, Government Printer, Pretoria.

Department of Transport (1998c), Working Documents for National Land Transport Bills and Cross-Border Transport Bill, unpublished, Pretoria.

Department of Transport (1999), *Moving South Africa - An Action Programme*, Government Printer, Pretoria.

Hemson, D. (2000), 'Policy and practice in water and sanitation', *Indicator SA*, vol. 17 (4), pp. 8-53.

Human Sciences Research Council (1999), *Results of November 1999 National Survey*, Pretoria.

Human Sciences Research Council (2000), *Results of September 2000 National Survey*, Pretoria.

Khosa, M.M. (1995), 'Transport and Popular Struggles in South Africa', *Antipode*, vol. 27, pp. 167-88.

Khosa, M.M. (1998), '"The Travail of Travelling": Urban Public Transportation in South Africa', *Transport Reviews*, vol. 18 (1), pp. 17-33.

Khosa, M.M. (2000a), *Empowerment through Service Delivery*, HSRC Publishers, Pretoria.

Khosa, M.M. (2000b), *Infrastructure Mandates for Change, 1994-1999*, HSRC Publishers, Pretoria.

Khosa, M.M. (2001), 'Facts, Fiction and Fabrication in South Africa: Service Delivery under Nelson Mandela', paper presented at the Association of American Geographers (AAG) Annual Conference, 28 February to 3 March, New York, USA.

Maharaj, M. (1999), Launch of Moving South Africa - The Action Agenda, Minister of Transport, 13 May.

Mail & Guardian (weekly newspaper), Johannesburg.

NTTT (1996), *Final Report of the National Taxi Task Team*, Department of Transport, Pretoria.

National Transport Policy Forum (1994), *Final Report of the National Transport Policy Forum*, Johannesburg.

Ndebele, S. (2000), KwaZulu-Natal Transport Budget Speech, 20 March.

Omar, D. (2000a), Address of the Minister of Transport to the National Council of Provinces on the National Land Transport Transition Bill, 13 April.

Omar, D. (2000b), Opening address by the Minister of Transport to the South African Transport Conference, 17 July 2000.

SAIRR (1998), *Annual Survey of Race Relations*, South African Institute of Race Relations, Johannesburg.

South Africa (1995), 'Urban Development Strategy of the Government of National Unity', *Government Gazette* 16679, Pretoria.

South Africa (1995), 'Rural Development Strategy of the Government of National Unity', *Government Gazette* 16679, Pretoria.

South Africa (1996a), *Green Paper on National Transport Policy*, Government Printer, Pretoria.

South Africa (1996b), *White Paper on National Transport Policy*, Government Printer Pretoria.

Part 2
Neighbouring States

9 Namibia's Economy: From Colonial Chattel to Postcolonial Pragmatism

DAVID SIMON

The Transition to Independence

After more than a century of colonial rule by Germany (1886-1915) and then South Africa, Namibia finally gained its independence on 21 March 1990. The intensity and nature of colonisation by its powerful neighbour, rather than by some distant European power, go a long way to explaining why Namibia had one of the longest liberation struggles in Africa, beginning in late 1968. Indeed, South Africa had long sought to incorporate the territory it inherited in 1919 as a C-class Mandate under the League of Nations system, as a *de facto* fifth province of the Union and then Republic. Only in 1977, following complex international negotiations, did the South African government accept the principle of independence for Namibia, although it took almost another thirteen years to achieve this under a United Nations-supervised transition (Simon, 1991; Wood, 1991).

President Sam Nujoma's ruling South West Africa People's Organisation (SWAPO) government confounded many observers by its pragmatism, keeping the economy on a generally sound footing with modest growth, and maintaining broadly co-operative relations with South Africa from the outset. Namibia has promoted a liberal, mixed economic system which encourages foreign direct investment and hopefully employment growth, while simultaneously seeking to redress the most oppressive and divisive legacies of apartheid. Since the late 1990s, however, the government has gradually become somewhat autocratic and unresponsive. Since 1996, overall economic growth has been slower than population growth, averaging 2.7 per annum. The first five full years of independence saw annual average growth of 5 per cent, enabling an increase in per capita incomes. The 5 per cent figure predicted for 2000 by the Economist Intelligence Unit failed to materialise as a result of lower than anticipated undersea diamond production; the provisional figure is 3.9 per cent.

External factors such as the end of the Cold War; the widespread abandonment of various forms of socialism around the world; the relatively unchallenged international agendas of economic liberalisation and privatisation; and aid conditionalities for continued support from key bilateral and multilateral donors, steered SWAPO away from Marxist socialism and state ownership of the key means of production. Internally, the new leadership's transition from exiled liberation movement to legitimate national government, and their pragmatic decision to avoid alienating the white business and farming elites in order to ensure economic stability, favoured a relatively hands-off approach.

This approach won plaudits from domestic and international business interests, the major international financial and multilateral agencies and bilateral Northern donors. However, many black Namibians expressed concern that the Foreign Investment Act of 1990 - symbolically one of the first new laws to reach the statute book after independence - introduced an investment regime which would reduce the benefits to Namibia by granting foreign firms too many incentives and would also have the effect of inhibiting the establishment and growth of local firms. Similarly, the 1992 Labour Act was perceived as being far less pro-worker than the trade union movement (including the SWAPO-affiliated National Union of Namibian Workers) had wanted or expected (Bauer, 1998). Notwithstanding such economic policies, unemployment has remained high, particularly in rural areas and among ex-combatants in SWAPO's military wing, PLAN. The loss of over 2,000 jobs with the effective closure in 1998 of Tsumeb copper and base metal mine, one of the country's largest employers, exacerbated the situation. The mine itself is exhausted, but the copper smelter and two other copper mines at Kombat and Otjihase eventually reopened under new management in late 2000. Australian bidders lost out to a consortium of the former mine managers and the Mineworkers' Union of Namibia under the name Ongopolo, which is employing roughly 1,000 people across the three sites (EIU, 2001).

Development brigades were established in 1993 in an effort to provide livelihoods for substantial numbers of ex-combatants in the northern communal areas, but the brigades' record has been problematic. Occasional protests, including street demonstrations in Windhoek city centre in October 1995 and 1997, subsequently drew attention to the ongoing severity of high unemployment among this core group of SWAPO members (EIU, 1995, 1996; *The Namibian*, 10 and 13 October 1995, 2 and 10 October 1997).

On the political front, efforts to address the apartheid legacy of divided administrative structures and institutions began immediately. One key task was the restructuring and integration of the civil service, government ministries and regional and local government institutions. Hitherto, these had been constituted along ethnic/racial lines in terms of apartheid and then neo-apartheid principles (Simon, 1991). In addition, a few new ministries or departments within ministries had to be established to handle functions previously controlled directly from South Africa, e.g. foreign affairs; national security and intelligence; customs and excise; maritime shipping and fisheries control; transport regulation and policy. From 1993, some rationalisation and restructuring occurred under a civil service review designed to reduce the size of the civil service.

Basic Economic, Social and Political Features

Before considering some of the principal economic policy issues facing Namibia since independence, it is important to outline a series of fundamental economic, social and political parameters and features that should inform any policy or planning process.

Economic and Social Parameters

- *Economic structure*: Namibia retains an open but narrowly based, export-oriented economy centred on the agricultural, fishing and minerals sectors. Limited diversification has occurred since independence, with particular efforts directed at expanding the manufacturing sector, but the economy remains very vulnerable to marked fluctuations in environmental/agricultural conditions and in world market prices for its base metal exports and oil imports in particular. Government services are the largest single contributor to gross domestic product (GDP), at around 25 per cent each year since shortly after independence; the financial services sector and real estate has also grown in recent years, to almost 10 per cent (Table 9.1). The table also reveals that, despite its traditional importance to GDP and employment in both the commercial and communal sectors, agriculture now contributes only 6-8 per cent of real GDP annually.
- *Environmental constraints*: Much of the country comprises desert (the coastal Namib Desert) and semi-desert, especially in the southern regions of Karas and Hardap as well as much of the Kunene region in the north-west and parts of the Omaheke in the east (Figure 9.1). Even the principal commercial cattle and game ranching areas of central-northern Namibia receive only 350-500 mm. of rainfall annually. Only the north-eastern parts of the Okavango region and Caprivi region are better watered, with a maximum of about 750 mm., thus making

dryland agriculture more secure. Historically controlled by whites, commercial agriculture is dominated by livestock rearing, especially beef cattle and karakul sheep (for lamb pelts). Cultivation has depended on water availability, and has thus been limited to the Hardap Dam scheme near Mariental, around artesian wells in certain parts of the south, and especially the Karstveld area between Otavi, Tsumeb and Grootfontein, where irrigated cropping of maize is centred. Elsewhere, production is mainly for subsistence in the communal areas, located in the former bantustans. The staple crops here are sorghum, pearl millet (*omahangu*) and maize. Some significant post-independence developments are outlined in the next section.

- *A small, unevenly distributed population*: The first post-independence census, undertaken in 1991, yielded a total population figure of 1.41m., somewhat less than previous UN and SWAPO estimates. Given the country's large surface area of some 826,000 km², this equates to 1.7 persons per km², although this national average conceals wide variations, from virtually zero in the Namib Desert to over 350/km² in densely settled parts of the Oshana and Ohangwena regions. Official 1998 estimates put the total population at 1.8m., implying a 3.2 per cent annual growth rate. Approximately 42 per cent of the population are under 15 years old, an age structure representative of countries with low levels of human development.

- *Profound income inequalities*: Although Namibia had a gross domestic product of US$3.04bn. in 1999, giving a figure of $1,820 per head, this obscures a deep-seated structural inequality in income distribution between the white minority of some 75,000 and the small but growing black elite on the one hand, and the majority of ordinary Namibians on the other. Without this skewing effect, Namibia would be classified unambiguously as a low-income rather than a middle-income country in the World Bank's ranking. As in South Africa, one of the most problematic legacies of apartheid has been the entrenchment through institutionalised means of a racial/ethnic hierarchy of status, incomes and state expenditures. Namibia's Human Development Index in 1998 was 0.632, i.e. within the medium human development category, but the difference between its GDP rank and HDI rank is -40 (UNDP, 2000), meaning that Namibia scores considerably higher on income than on social aspects of development. This legacy will take considerable time to redress, despite the rapid equalisation of pensions, public sector wages and per capita expenditures on health and education services after independence.

Table 9.1   Namibia: Gross Domestic Product by Sector, 1995-1999
(N$m. unless otherwise indicated; constant 1990 prices)

	1995	1996	1997	1998	1999[a]
Agriculture[b]	521	573	509	477	493
% change, year on year	-1.8	9.9	-11.1	-6.4	3.3
% of total	7.6	8.2	7.1	6.5	6.5
Fishing	307	312	306	370	386
% change, year on year	7.5	1.6	-2.0	20.9	4.4
% of total	4.5	4.5	4.3	5.0	5.1
Mining & quarrying	1,291	1,339	1,391	1,342	1,388
% change, year on year	5.2	3.8	3.9	-3.5	3.4
% of total	18.9	19.2	19.4	18.3	18.4
of which:					
Diamond mining	904	940	939	955	1,042
% change, year on year	7.1	3.9	-0.1	1.7	9.2
% of total	13.2	13.5	13.1	13.0	13.8
Manufacturing[c]	922	856	972	1,008	1,001
% change, year on year	0.6	-7.2	13.6	3.7	-0.7
% of total	13.5	12.3	13.5	13.7	13.3
of which:					
Fish processing	238	145	228	243	228
% change, year on year	-7.8	-39.1	56.9	6.9	-6.2
% of total	3.5	2.1	3.2	3.3	3.0
Construction	221	238	175	153	157
% change, year on year	2.6	7.8	-26.5	-12.7	2.7
% of total	3.2	3.4	2.4	2.1	2.1
Electricity & water	103	85	56	76	93
% change, year on year	24.5	-17.4	-34.7	35.9	22.8
% of total	1.5	1.2	0.8	1.0	1.2
Wholesale & retail trade	478	487	515	508	531
% change, year on year	4.5	1.8	5.7	-1.3	4.6
% of total	7.0	7.0	7.2	6.9	7.0
Hotels & restaurants	128	120	151	171	174
% change, year on year	13.7	-6.2	25.6	13.1	1.9
% of total	1.9	1.7	2.1	2.3	2.3
Transport & communications	394	422	469	496	509
% change, year on year	10.6	7.1	11.0	5.7	2.7
% of total	5.8	6.1	6.5	6.7	6.8
Finance & real estate	562	600	644	680	680
% change, year on year	2.2	6.7	7.4	5.6	-0.1
% of total	8.2	8.6	9.0	9.3	9.0
Government services	1,685	1,717	1,768	1,846	1,896
% change, year on year	-0.1	1.9	3.0	4.4	2.7
% of total	24.7	24.6	24.6	25.1	25.2
Social & community services	65	66	65	63	63
% change, year on year	-0.7	1.3	-1.4	-2.6	-0.3
% of total	1.0	0.9	0.9	0.9	0.8
Other producers	153	156	159	161	163
% change, year on year	2.0	2.0	1.6	1.8	1.0
% of total	2.2	2.2	2.2	2.2	2.2
GDP at factor cost	**6,832**	**6,973**	**7,181**	**7,351**	**7,534**

Notes:

a Preliminary.
b Agricultural coverage has been expanded from the main livestock and crop farming activities to include production of milk, eggs, sunflower, cotton, ground nut and lucerne crops, and the value of the ostrich stock.
c The contribution of other manufacturing value added was raised substantially by incorporating results from a survey of non-meat and fish-processing activities conducted in 1996.
Source: EIU, 2000.

Indeed, class and social differentiation have probably increased (Tapscott, 1993). This factor intersects with three other features that Namibia shares with several of South Africa's neighbours, namely a geographically uneven resource endowment, diverse prevailing environmental conditions, and the legacy of the colonial political economy (especially in terms of the location of urban centres, transport infrastructure, and the institutionalised distinction between commercial and communal lands), to produce pronounced regional inequalities of income and income-earning opportunities. Addressing these is not proving easy.

- *Dependence on South Africa*: At independence, there was real concern that a hostile South Africa would seek to destabilise the new state. This would have been remarkably easy in view of the disparity in their economic strengths, South Africa's longstanding occupation, and the high Namibian dependence upon South Africa that had been consciously fostered during the long decades of segregation and then apartheid. South African companies dominated diamond and some other mineral mining, as well as the retail, hotel, banking and financial services sectors. The transport systems had been integrated until the late 1980s, something that, coupled with South Africa's continued occupation of the coastal enclave of Walvis Bay, provided the Pretoria government with the ability to strangle Namibian trade, i.e. using transport as a noose rather than a lifeline (Simon, 1989, 1998).

Political Parameters

Given Namibia's particular geopolitical and politico-economic history, it is hardly surprising that several clear domestic political tensions have emerged during the first eleven years of statehood. Although generic to new postcolonial states, they have been given a particular resonance in the context of the overlapping economic, social and environmental conditions outlined above.

Figure 9.1 Namibia: Regional and Local Government

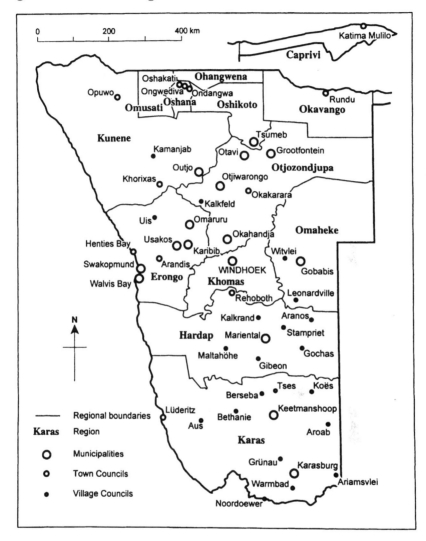

- *Centralisation versus decentralisation*: Namibia is a unitary state divided into thirteen regions (Figure 9.1). The regional councils have been given substantial administrative and developmental powers and responsibilities. However, partly on account of the influence of a strong centralising tendency within SWAPO, the resources provided to the regions are wholly inadequate to fulfil this mandate under the Regional Councils Act of 1992. This has precipitated a crisis of credibility as a result of the councils' generally weak performance, and arguments have raged over the last few years as to whether the councils should be abolished or the extent of decentralisation be increased (Simon 1996a; Forrest 1998). Voter apathy – turnout in the 1998 regional and local council elections was a miserly 30 per cent – has been one factor behind the government's recent apparent intention to promote rather than restrict decentralisation.

- *Consolidation of democracy in the state and within SWAPO*: The multi-party system has become well established, albeit with SWAPO emerging as the dominant party. The Democratic Turnhalle Alliance (DTA), which was substantially funded by South Africa before independence in an effort to forestall a SWAPO victory, has lost credibility and much ground. However, SWAPO's dominance of domestic politics, combined with its long authoritarian tradition as a largely exiled liberation movement, has enabled it to become increasingly centralist and unaccountable internally as well as to the country at large. Both President Sam Nujoma's personal style of reconciliation and the feel-good factor following the peaceful resolution of two inherited boundary disputes with South Africa (see below) have waned over time. The subsequent boundary dispute with Bostwana over Kasikili/Sedudu Island in the Chobe River, failed to generate anything like the same passions inside Namibia. The Congress of Democrats, established in early 1999, emerged as the principal challenger to SWAPO's dominance, although it gained a somewhat disappointing 9.9 per cent of the vote in the December 1999 general election. This was still adequate to become the official opposition, however, since the DTA gained only 9.4 per cent (Bauer, 1999; Oxford Analytica, 1999b; Simon, 2000a).
- *Emerging challenges to nationhood*: In 1998 and 1999, the first two substantive postcolonial challenges to the boundaries and nature of the young state emerged, fuelled by dissatisfaction over the issues of centralisation and the lack of accountability and representation just outlined. The government's secret decision to send Namibian troops to fight in support of Laurent Kabila's government in the Democratic Republic of Congo (DRC) in 1998 was widely condemned both within and beyond parliament. This was seen as an abuse of power, which severely dented the government's domestic credibility and somewhat tarnished the country's international standing. Second, an opposition-led secessionist uprising occurred in the Caprivi Region in August 1999. A significant grouping of Caprivians have long sought independence separately; their senses of identity are more bound up with kin and co-linguists in northern Botswana and south-western

Zambia. The SWAPO government's handling of these issues in the context of the wider problems, coupled with the collapse of the DTA, within which the key players opposed to SWAPO had been prominent, precipitated an ill-judged uprising in which some fifteen people died. It was put down firmly by the Namibian Defence Force and police, with a considerable level of unnecessary brutality. Over 2,600 people fled to Botswana and were treated as refugees by the UNHCR; within a year, most had returned home under a Namibian promise of non-persecution. The immediate crisis may be over, but the resentment and underlying issues have not gone away (*Business Day*, 1999; *The Cape Times*, 1999; *Oxford Analytica*, 1999a; *The Namibian*, 1999).

Economic Challenges and Achievements since Independence

Sovereignty over Strategic Resources and Infrastructure

The two above-mentioned inherited boundary disputes with South Africa held both geopolitical and major economic importance. The former revolved around territorial integrity and the legitimation of sovereign nationhood as propounded by SWAPO and the international community. However, no less important was the securing of key economic resources. The disputes centred on the precise demarcation of the mutual border along the Orange River and especially at its mouth, and on the status of the Walvis Bay enclave then still occupied by South Africa, which claimed legitimacy on the basis of a nineteenth-century British cession thereof to the then Cape Colony. Both issues were long-running but had been deferred by the UN until after Namibian independence so as not to jeopardise that objective (Evans, 1990; Simon, 1991, 1996b).

Despite the huge historical baggage and Namibian concerns that apartheid South Africa was still secretly planning to undermine the new SWAPO government and Namibian independence, joint technical commissions were established and both disputes were resolved peacefully. The Orange River boundary was quickly agreed as the mid-point of the river, in accordance with international law, whereas South Africa had claimed everything up to the river's northern (i.e. Namibian) bank. The Walvis Bay issue took rather longer to finalise because of the perceived sensitivities of the mainly South African white personnel stationed there, as well as a feared backlash among South Africa's white electorate. Following a period of joint administration from late 1992, the enclave and its associated offshore islands were reintegrated into Namibia at midnight on

28th February 1994 in an emotion-charged ceremony which evoked
memories of Namibian independence itself. The difference in terminology
between the two sides is also indicative of their respective positions: the
South African government referred to the process as a 'handover' while
Namibia has always talked of 'reintegration' (Simon, 1996b, 1998). The
economic importance to Namibia of these disputed settlements is profound
in location-specific as well as more general terms.

The Orange River and the diamond industry: Namibia now has rights to a
key water resource at its southern extremity in what is a harsh desert
environment. This is used for human consumption, in certain mining
operations, and has enabled the development since 1991 of a major
commercial farming enterprise at Aussenkehr near Noordoewer, geared
principally to the export to Europe of top-grade grapes and melons. These
ripen ahead of rival produce from South Africa and other southern
hemisphere producers, thus gaining premium prices abroad. Production has
now expanded to enable some also to be marketed domestically and in
South Africa.

Equally important, however, is Namibian sovereignty over several
hundred metres of rich diamond-bearing alluvium at the river mouth and on
the continental shelf. Recovery of underwater gems using new dredging
and suction technology is the fastest-growing part of the diamond industry,
as older onshore deposits approach exhaustion, and is providing both
employment and crucial tax revenues. In 1999, offshore diamond
recoveries by Debmarine, Namdeb's subsidiary, and the handful of
specialised small companies reached 56 per cent of the country's total
output of 1.639 million carats (EIU, 2000).

In this context, it is worth mentioning that in 1994 the Namibian
government took a 50 per cent stake in Consolidated Diamond Mines
(CDM), hitherto owned by South Africa's De Beers company, the world's
largest diamond producer. This joint venture became the Namdeb
(Namibian De Beers) Diamond Corporation. This was a remarkably
favourable deal for the government, to forestall the threat of nationalisation.
The government justified its move as a key element of the process of
securing sovereignty over one of the leading mineral resources, from which
it had hitherto received inadequate tax revenue on account of alleged
transfer pricing by CDM/De Beers.

Walvis Bay: Walvis Bay town is of crucial importance on account of its
efficient harbour, which is both the headquarters of Namibia's pelagic and
deepwater fishing industry and the principal port through which virtually
all the seaborne imports and exports pass. Gaining control over these

facilities on generally reasonable terms has been central to the economic sovereignty argument in infrastructural as well as resource terms. Finalisation of the deal to acquire the physical infrastructure in the harbour proved more problematic than the preceding political accommodation, because ownership was vested mainly in various commercialised South African parastatal corporations, especially Transnet and its Portnet subsidiary, rather than in the state itself. The financial transaction eventually took place in December 1995, with a formal handover ceremony for the equipment on 22 March 1996, six years and one day after independence and eighteen months after the formal reintegration of Walvis Bay (Simon, 1998).

Since 1993/4, the harbour has handled between 1.81m. and 1.91m. tonnes of trade annually, approaching its current capacity of some 2m. tonnes, although expansion and upgrading work is in progress. Petroleum, fish products, wheat, maize and coal are (in descending order) the principal commodities landed, whereas salt, manganese ore, fish products, copper and lead, and fluorspar are the main exports (Simon, 1998; EIU, 2000). The reintegration of Walvis Bay necessitated the establishment of a Namibian Ports Corporation (Namport) to regulate and operate the port sector and associated infrastructure. In terms of current international conventional neoliberal wisdom, there is an intention to separate these two functions at an appropriate future date.

Fisheries: Walvis Bay's reintegration also brought the entire Namibian fishing fleet under local control and has greatly facilitated the tasks of fisheries management. Overfishing of the pelagic stocks almost wiped out the industry in the 1960s and 1970s. A gradual recovery until independence was accompanied by great expansion of the demersal fisheries for white fish (especially hake), and within the last decade also deepwater species like the slow-breeding orange roughy. However, conservation and monitoring have been greatly enhanced since independence and the reintegration of Walvis Bay.

The country's Exclusive Economic Zone (EEZ) has been patrolled quite effectively, and several Spanish trawlers and their catches were confiscated for illegal fishing in landmark court cases shortly after independence. The authorities set total allowable catches (TACs) for each species on an annual basis in the light of fish stock monitoring and prevailing conditions. Most stocks had recovered well during 1997-99 following the disastrous El Niño event that had reversed the strong catch increases from 1991-93. However, another anticipated downturn was reflected in much lower TACs for 2000 compared with 1999 (Table 9.2).

Table 9.2 Total Allowable Catches for Namibia's Major Catch Species (tons)

Species	1999	2000	% change
Pilchards	55,000	15,000	-72.7
Hake	210,000	194,000	- 7.7
Horse mackerel	375,000	410,000	+ 9.3
Orange roughy	6,000	2,100	-60.0

Source: EIU, 2000, p. 32.

Economic Diversification and Energy Self-reliance

Namibia's circumstances preclude a rapid diversification and industrialisation programme akin to that adopted in many other newly independent states in earlier decades. Pragmatism fortunately prevailed, and the emphasis has been on selective and generally appropriate targets, combined with enhancing output in existing activities and broadening the tax base. Diversification within the diamond and fishing industries was discussed above. Increased fish processing at Walvis Bay, in particular, as a result of higher catches has contributed to employment growth. However, these employees are subject to periodic lay-offs as and when TACs are reduced.

Subsidies to commercial agriculture have been reduced while access to technical assistance and credit by farmers in the communal areas has been facilitated. Hence, Namibia was quickly able to reduce the extent of its dependence on imported cereals and other foods, through a combination of expanded commercial irrigation in the Karstveld and enhanced production and marketing in the northern communal areas. However, Namibia is unlikely ever to achieve self-sufficiency. Efforts to integrate the agricultural sector and to reduce the inherited dichotomy between commercial and communal areas and farming systems, have borne some fruit. Land reform efforts have been slow, market-led and based on the resettlement of landless peasants and ex-combatants, as well as encouraging large-scale ranchers in the communal areas to relocate on to commercial farms. New vested interests in land among the ruling elite have been a key handicap to progress, but recent events in Zimbabwe have spurred renewed commitments to redistribution in order to prevent similar actions in Namibia, which the government opposes.

Industrial Diversification

The principal focus of industrial and economic diversification strategies, beyond increasing the output of existing activities, has been to add value to existing activities through processing prior to export (i.e. beneficiation). Since the mid-1980s, CDM/Namdeb has undertaken diamond sorting in Windhoek. The cutting and polishing of precious and semi-precious stones is another sphere in which Namibia has a natural comparative advantage and some progress has been made (Currey and Stoneman, 1993; EIU, 2000).

Also noteworthy was the establishment of an export processing zone (EPZ) at Walvis Bay, seen grandiosely by some as part of an envisaged major transport corridor linking the port to the central African interior. Its record is not impressive, and several companies quickly closed down. Like many EPZs worldwide, this became rather an enclave with little organic linkage to the rest of the country. The principal apparent benefit is therefore in terms of new jobs of variable quality. The 1995 Export Processing Zones Act has enabled any factory within Namibia to apply for EPZ status subject to certain conditions. This has undermined whatever previous locational advantage the Walvis Bay EPZ initially had. Moreover, the government's suspension of key elements of the Labour Act within EPZs enraged trade unions (Simon, 1998, pp. 119-21).

On the other hand, initial investment plans by Citroën and Saab-Scania to locate car and truck assembly plants at Gobabis and Witvlei respectively, but catering to the South African and SADC market as a whole (Simon, 1991, pp. 23-4), came to nothing on account of transport costs, logistics and stronger incentives elsewhere. Very recently, though, the Malaysian textile firm, Ramatex, cited more generous incentives as its reason for opting to build a R1bn. factory in Namibia rather than South Africa's Eastern Cape (*Business Day*, 2001).

Energy and Electricity Supply

Namibia's government has also sought to address the country's dependence on energy imports for all of its petroleum and coal and about half of its electricity requirements. Consequently, the Namibian government has vigorously promoted a joint venture with Angola to construct a second large dam at Epupa Falls on the Kunene River. This has proved highly controversial both locally and internationally, on account of the $543m. price tag in 1998 terms, the environmental and social impact upon the local OvaHimba communities, and for reasons common to most large dam schemes (McCulley, 1998; Simon, 2000b; World Commission on Dams, 2000). As a result, Nordic funding beyond the feasibility stage has been withdrawn and the World Bank is also unlikely to support it (Oxford Analytica, 1999a). This has almost certainly scuppered the plan.

An alternative solution lies in development of the promising offshore Kudu gas field. The intention is to construct a 750 megawatt gas-fired power station at Lüderitz to meet Namibian demand for electricity, and to export gas to Cape Town by pipeline for a new power station to be built there. The cost of electricity from Kudu was estimated in 1998 to be 20 per cent cheaper than from Epupa. Either project would create Namibian self-sufficiency in electricity (ibid., 1999a).

Conclusions

Independent Namibia has pursued a generally pragmatic economic policy, with a liberal foreign investment regime and an incentive-led EPZ strategy to promote job creation and industrial diversification. Only modest progress has been made on this score, as with efforts at self-sufficiency in electricity supply. Perhaps the most significant successes since 1990 have been the resolution of boundary disputes with South Africa, thereby establishing Namibian sovereignty over key coastal resources and infrastructure and enhancing the substance of political independence beyond the largely symbolic. Associated with the settlement of the Orange River and Walvis Bay disputes has been a major boost to offshore diamond recovery and to more sustainable fisheries exploitation within Namibia's EEZ. Domestic cereal production has also increased markedly.

Although Namibia has enjoyed general stability since independence, the resources and political will to promote substantive restructuring to overcome the inherited legacies of apartheid and South African rule have not been adequate. Unemployment and poverty remain widespread and may even have risen in relative terms. Frustrations are therefore high outside the ranks of the elite and middle classes.

References

Bauer, G. (1998), *Labor and Democracy in Namibia 1971-1996*, James Currey, Oxford; David Philip, Cape Town; Ohio University Press, Athens.
Bauer, G. (1999), Challenges to Democratic Consolidation in Namibia, in Joseph, R. (ed), *State, Conflict, and Democracy in Africa*, Rienner, Boulder and London, pp. 429-48.
Business Day (1999), 3 and 6 August, Johannesburg.
Business Day (2001), 27 June, Johannesburg.
Cape Times, The (1999), 3 and 5 August, Cape Town.
Currey, S., and Stoneman, C. (1993), 'Problems of Industrial Development and Market Integration in Namibia', *Journal of Southern African Studies*, vol. 19, pp. 40-59.

EIU (1995), *Country Report: Namibia, Fourth Quarter 1995*, Economist Intelligence Unit, London.

EIU (1996), *Country Profile 1995-96: Namibia*, Economist Intelligence Unit, London.

EIU (2000), *Namibia, Swaziland: Country Profile 2000*, Economist Intelligence Unit, London.

EIU (2001), *Country Report: Namibia, First Quarter 2001*, Economist Intelligence Unit, London.

Evans, G. (1990), 'Walvis Bay: South Africa, Namibia and the Question of Sovereignty', *International Affairs*, vol. 66, pp. 559-68.

Forrest, J.B. (1998), *Namibia's Post-Apartheid Regional Institutions: the Founding Year*, University of Rochester Press, Rochester, New York.

McCulley, P. (1998), *Silenced Rivers: the Ecology and Politics of Large Dams*, Zed, London.

Namibian, The (1995), 10 and 13 October, Windhoek.

Namibian, The (1997), 2 and 10 October, Windhoek.

Namibian, The (1999), daily reports in August and September, Windhoek.

Oxford Analytica (1999a), ' Namibia: Energy Development', *Oxford Analytica Daily Brief*, vol. 5, pp. 14-16.

Oxford Analytica (1999b), 'Namibia: SWAPO entrenched', *Oxford Analytica Daily Brief*, vol. 5, pp. 15-17.

Simon, D. (1989), 'Transport and Development in Namibia: Noose or Lifeline?', *Third World Planning Review*, vol. 11, pp. 5-21.

Simon, D. (1991), *Independent Namibia One Year On*, Conflict Studies 239, Research Institute for the Study of Conflict and Terrorism, London.

Simon, D. (1996a), 'What's in a Map? Regional Restructuring and the State in Independent Namibia', *Regional Development Dialogue*, vol. 17, pp. 1-31.

Simon, D. (1996b), 'Strategic Territory and Territorial Strategy: the Geopolitics of Walvis Bay's Reintegration into Namibia', *Political Geography*, vol. 15, pp. 193-219.

Simon, D. (1998), 'Desert Enclave to Regional Gateway? Walvis Bay's Reintegration into Namibia', in Simon, D. (ed), *South Africa in Southern Africa: Reconfiguring the Region*, James Currey, Oxford; David Philip, Cape Town; Ohio University Press, Athens, pp. 103-28.

Simon, D. (2000a), 'Namibian Elections: SWAPO Consolidates its Hold on Power', *Review of African Political Economy*, vol. 27, pp. 113-15.

Simon, D. (2000b), 'Viewpoint: Damm(n)ed Development?', *Third World Planning Review*, vol. 22, pp. iii-ix.

Tapscott, C. (1993), 'National Reconciliation, Social Equity and Class Formation in Independent Namibia', *Journal of Southern African Studies*, vol. 19, pp. 29-39.

United Nations Development Programme (2000), *Human Development Report 2000*, Oxford University Press, New York (available at http://www.undp.org/hdro).

Wood, B. (1991), 'Preventing the Vacuum: Determinants of the Namibia Settlement', *Journal of Southern African Studies*, vol. 17, pp. 742-63.

World Commission on Dams (2000), *Dams and Development; a New Framework for Decision-Making*, Earthscan, London and Sterling, Virginia.

10 Adjustments to Globalisation: The Changing Economic Geography of Botswana

AGNES MUSYOKI AND MICHAEL BERNARD KWESI DARKOH

Introduction

The process of globalisation is rapidly influencing and changing the economic geography of southern Africa. The forces of globalisation linked to structural adjustment programmes and market liberalisation measures are triggering shifts in the economies and organisation of many developing countries. In addition, there is occurring the formation of new alliances, the information explosion, development of new ideas and abandonment of old beliefs, and increasing reliance on science and technology (De Blij and Muller 2000). Many leaders of African nations are calling for a 'renaissance' as a way out of the economic stagnation that afflicts much of the continent. This chapter discusses the changing economic geography of Botswana within the context of the Southern African Development Community (SADC) and the wider context of globalisation. It highlights a set of adjustments made in response to globalisation and the resultant consequences for the making and re-making of economic geographies in Botswana.

General Characteristics of the Botswana Economy

The economy of Botswana has grown rapidly during the last thirty years. At independence in 1996 Botswana was one of the poorest countries in the world. The economy has since grown at such a rapid rate that today Botswana is classified by the World Bank as a middle-income country. For much of the post-independence period growth in GDP is unparalleled anywhere in modern economic history. Real per capita income has grown in this period at 7 per cent per annum. What has surprised many analysts is the fact that this tremendous growth has been led by mineral wealth. Many countries have failed to prosper despite discovery of minerals because of the so-called 'Dutch disease' where the rapid development of one sector, in this case the mineral sector with its substantial revenues to government, has knock-on effects that serve to retard development elsewhere.

Most of Botswana's success is related to prudent policy formulation and implementation and the sustenance of good fiscal and macro-economic policies. The country is also investor-friendly. In January 2001 *African Business* ranked Botswana second to Singapore in terms of best emerging market risk, ahead of Taiwan, Chile and even the United Arab Emirates. The indicators used to determine investor confidence included state of finances, political risk and business environment factors. In all of these indicators Botswana, according to the report, performed exceptionally well. As compared with Botswana, other African countries, including Mauritius, ranked low, while Namibia, South Africa and Tunisia were rated moderate to low risk.

National development objectives seek to improve the well being of all Botswana. Development is seen in the broader context of political and social welfare as embodied in Botswana's four national principles of democracy, development, self-reliance and unity. In planning national development, Botswana has been guided by the cardinal objectives of sustained development, rapid economic growth, economic independence, and social justice. To manage the economy, Botswana has adopted a broad investment policy that takes into account the importance of safeguarding and maintaining its foreign exchange reserves, ensuring their liquidity if needed, and earning the best returns possible within acceptable limits of risk. One of the primary reasons that Botswana left the Rand Monetary Area and introduced the Pula was to enable the country to accumulate foreign exchange reserves, the earning on which could be a source of additional revenues. Little was it anticipated that foreign exchange would accumulate to over P13bn. by the end of 1995 and P23bn. by 2000. Recently, earnings on foreign exchange reserves have become the second largest source of government revenue. Since independence, Botswana has endeavoured to maintain a stable macro-economic environment, encouraging savings and capital accumulation as well as facilitating long-term planning and investment decisions. In 1990, Botswana adopted the Financial Sector Development Strategy. This strategy emphasises the maintenance of positive real interest rates as the key intermediate objective of monetary policy and this has helped stimulate savings.

Despite the sound macro-economic and fiscal policies pursued by Botswana, progress towards the four national objectives has been uneven. The discovery and production of diamonds has stimulated rapid growth, as measured by the nation's GDP, although there has been only a slow process of re-investment of the diamond proceeds in domestic productive activities which translate into income-earning opportunities for those seeking work. The diversification of the economy which is necessary for the promotion of social justice, economic independence and sustained development has proved to be relatively difficult. Undoubtedly, the sound macro-economic policies pursued by Botswana have provided a good climate for growth and prosperity. This positive trend when analysed at the sector level shows that Botswana could perform better if appropriate measures are put in place. The following sections examine the performance of each of the key sectors in Botswana's economy.

The Agricultural Sector

Botswana's rapid economic changes have been characterised by a decline in the agricultural sector. At independence Botswana depended solely on agriculture which employed 90 per cent of the people in the labour market. At the same time it contributed about 43 per cent to the GDP. A sharp decline followed so that by the 1990s the shares of GDP and employment had fallen to about 4 per cent and 16 per cent respectively. The decline has been associated with growth of the economy generally and of other sectors particularly mining. Drought and desertification also adversely affected growth in agriculture. The country generally does not have fertile soils to support large-scale agriculture. Cattle-farming is the only agricultural activity which has prospered in Botswana. The other sub-sectors in agriculture have not done well even when the government injected funds for support of emerging farmers.

In the 1990s farming was beset by many problems which included the outbreak of cattle lung disease (contagious bovine pleuropneumonia) leading to the slaughter of 312,000 cattle (12 per cent of the country's cattle population). What is significant, however, is that even when weather and other conditions were favourable, the agricultural sector in Botswana continued to decline, which indicates structural problems within the sector. A major concern within the agricultural sector is the dominance of traditional agriculture. If agriculture is to contribute to national income and exports, changes are inevitable. Several factors account for the poor performance of agriculture:

- lack of access to capital and expertise;
- poor management practices and production methods;
- absentee landlords and farmers who work in agriculture on a part-time basis;

- communal grazing tends to promote overgrazing and may promote spread of disease;
- Botswana's maintenance of multiple homesteads has negative impacts on farming;
- cattle are seen as a status symbol;
- very weak linkages between agriculture and the productive sectors;
- sex ratio, age distribution and HIV/AIDS related diseases;
- perception of agriculture as a low-paying occupation and desertion of agriculture by young people.

Land Ownership

Three major land tenure systems exist in Botswana. The first, freehold land giving perpetual and exclusive rights to the owners, occupies only five per cent of the total land area. This type of land tenure is mainly found close to the eastern and western borders in some of the most fertile lands in the country. The second, state lands, include national parks, game reserves, wildlife management areas and forest reserves covering 25 per cent of the land. The rest of the land is communally owned tribal land, giving land access to all Batswana.

In Botswana, traditional and modern tenure systems co-exist with minimum conflict. Land is usually allocated to both men and women by Land Boards and managed through traditional and local authority. The country's legislation recognises the rights of all Batswana to land. Despite this seeming gender neutrality other dynamics interplay to deny women full access to land and land rights. Upon marriage many women voluntarily register their land in the husbands' names and in the event of divorce or death of husband women often lose the right to land and do not contest their rights in court. This situation is due to patriarchal practices still prevalent in most of the country leading to higher levels of landlessness and poverty among women both in rural and urban areas (Kalabamu 1998; Musyoki, 1998).

The Livestock Sector

The second most important sector of Botswana's economy outside mining is livestock rearing. This activity has both economic and cultural value having been in existence since the fourth century. Major changes in the industry were witnessed in the colonial and post-colonial periods in Botswana. Some of the most significant changes include a marked increase in cattle population, an increase in the number of water points across the country, fencing of the country into separate veterinary regions, the commercialisation of the cattle industry and the opening of export markets, initially in South Africa, followed later by regional markets such as Zimbabwe, Zambia and Mauritius and finally the European Union.

The major catalyst for the expansion of the industry was the construction and provision of boreholes, which are to be found throughout the country. The Tribal Grazing Land Programme (TGLP) of 1975 encouraged big cattle owners to move to commercial ranches with the aim of reducing pressure on the grazing resources in the communal areas. However, the TGLP did not lead to the permanent removal of large herds from communal areas nor did it lead to the reduction of grazing pressures there. Instead, the managers of commercial ranches were allowed to exercise dual grazing rights, that is, simultaneously to use private pastures as well as communal areas, regardless of carrying capacity (Parson, 1981). This issue of dual grazing rights still plagues the pastoral economy and is one of the problems to be contended with if environmental sustainability is to be ensured in the rangelands of Botswana.

An important aspect of the cattle industry in Botswana is that it has not contributed as much to government revenue as expected, despite the fact that it is dominated by some of the wealthiest people and political elites in the country (Pearce, 1993). During the late 1970s, public finance of the cattle sector averaged P12.1m. (approximately $3m.) per year while flows from this sector to the government in the form of taxes and payments for services averaged P7.5-m. At independence in 1966 meat accounted for 97.2 per cent of the exports, but by 1987 it had fallen to only 3.1 per cent, a level that has remained little changed.

Livestock expansion in Botswana has generated several land use conflicts. These are most pronounced between the livestock and arable sectors on the one hand and the wildlife and tourist sectors on the other. The expansion of livestock into wildlife areas has reduced the amount of land used by wildlife for grazing. Even more pronounced is the effect of veterinary cordon fences. Such veterinary fences have tended to create an artificial barrier for migratory wildlife species which have continued to die in large numbers along them. The increasing decline of the country's wildlife population raises fears concerning the future development of Botswana's promising tourism economy.

As livestock production remains the backbone of the rural economy but continues to contribute poorly to the national economy, it becomes necessary for government to diversify the rural economy. Following concern over the general performance of the agricultural sector despite resource flows, the government commissioned a comprehensive review of the sector in 1988/89 in order to formulate policies and programmes that could enhance productivity, competitiveness and sustainability. The outcome of the sectoral review, which culminated in the adoption of the Agricultural Development Policy in 1991 by Parliament, strongly advocated the replacement of the food self-sufficiency policy strategy that had been in operation since independence with a food security policy. This policy aimed at an overall enhancement of the income levels of households especially in the rural areas. It targeted subsidies to farmers and promoted potentially viable agricultural and non-agricultural programmes, advocating that shortfalls in food requirements be met from commercial imports. Policy at both the household and national levels has emphasised the diversification of income sources to enable individuals and households to have access to food and other basic needs.

Diversification of Agriculture

In order to improve on the performance of agriculture under the new food security policy, several diversification efforts have been undertaken. The Financial Assistance Programme (FAP) (Box 10.1) has been the main source of capital for farmers. Local production of chickens, meat, eggs and milk has contributed to self-sufficiency in these commodities. Horticultural farming has, however, failed despite funds injected into it through the FAP programme.

Box 10.1 Financial Assistance Policy in Botswana

The Botswana Government has developed a package of incentives for national and international investors, to stimulate investment and create employment opportunities. The Financial Assistance Policy (FAP), which is supervised by the Ministry of Finance and Development Planning, is one of the main investment incentives available to entrepreneurs. FAP, which was first established in 1982, provides financial assistance to entrepreneurs in the industrial sector, as well as for certain agricultural and small mining activities. Grants are related to the number of citizen jobs created. Businesses may receive FAP support for: small-scale projects with fixed capital investment of up to P25,000; medium-scale projects with fixed capital investment between P25,000 and P900,000; and large-scale projects with fixed capital investment in excess of P900,000. During National Development Plan 6, nearly 1,200 projects received FAP financing, with the value of the project investments totalling over P250m., including both FAP and funds raised by investors directly.

Source: Ministry of Finance and Development Planning, *National Development Plan 7.*

Support for commercial farming has been provided for development of farms in Pandamatenga in northern Botswana. In most cases farmers have run into problems after an initial period of support. It is encouraging to note that Debswana is supporting agricultural projects even though performance is generally poor. More farmers are getting involved in game and ostrich farming as means of diversifying the agricultural base.

Mining

Mining is the most important sector in the Botswana economy. Indeed, the mining industry was largely responsible for the graduation of Botswana from recurrent budget aid from the British Government and entitlement to International Development Association resources in the early 1970s and from the world's least developed countries in the late 1980s (Gaolathe, 1997). Tables 10.1 and 10.2 depict the changing contribution of the minerals sector to Botswana's economy in terms of Gross Domestic Product (GDP) for the period 1972-95.

It must be appreciated that minerals have been exploited in order to make implements for hundreds of years. Indeed, gold was mined in Botswana as far back as the nineteenth century. Nevertheless, the major surge of the modern mining economy began in the 1960s. Indeed, mining received a new momentum after independence in 1966, which led to significant discoveries of diamonds, copper-nickel and coal which are the

Table 10.1   Contribution of Minerals to Botswana's Gross Domestic Product

Year/Indicator	GDP (Pula Million)	Mining GDP (Pula Million)	Mineral Share of GDP (%)
1972	103.6	11.2	11
1976	300.4	42.0	14
1980	875.5	201.6	23
1981	899.9	201.6	22
1983	1153.1	366.6	32
1985	1828.6	753.1	41
1987	2809.8	1229.8	44
1989	5836.8	2969.3	51
1991	7475.2	3012.0	40
1993	9126.0	3042.3	30
1995	12530.3	4086.3	33

Source: Central Statistics Office.

Table 10.2 Contribution of Minerals to Government Revenue in Botswana

Indicator/ Year	Government Mineral Revenues (Pula Million)	Total Government Revenues (Pula Million)	Mineral Share of Total revenues (%)
1972	1.0	19.3	5
1976	23.3	87.8	27
1980	76.6	249.1	31
1981	101.1	306.6	33
1983	99.5	393.7	25
1985	376.5	802.9	47
1987	845.0	1 547.5	55
1989	1 508.1	2 556.0	59
1991	2 005.3	3 740.7	54
1993	1 866.1	4 652.2	40
1995	2 278.7	4 492.5	51

Source: Ministry of Finance and Development Planning.

most important minerals. Overall, diamond mining represents the most important sub-sector whose contribution to the economy has been exceptional.

One result of the mining activities was the development of infrastructure and opening up of remote areas in the northern part of the country. However the most significant impact of mining, particularly diamond mining, has been its monetary contribution in the form of taxes and royalties for the national accounts. Botswana is one of the world's largest producers of diamonds. The diamond industry relies on De Beers and gems are sold through the Central Selling Organisation (CSO) which influences prices through cartels. The major diamond mines are found at Orapa, Letlhakane and Jwaneng (Figure 10.1). The Orapa mine is the largest in the world with a surface area of 110.6 hectares.

Copper-nickel production played a significant economic role until the 1970s when fluctuations in prices rendered the sub-sector uneconomical. This situation changed in the 1990s although the profitability of the mine has not improved and it continues to be heavily subsidised by government. This has raised concerns for the economic future of Selebi-Phikwe town. Special efforts are being made therefore to attract new economic activities into the town and ensure a livelihood and growth for this settlement of nearly 50,000 people.

Figure 10.1 Botswana

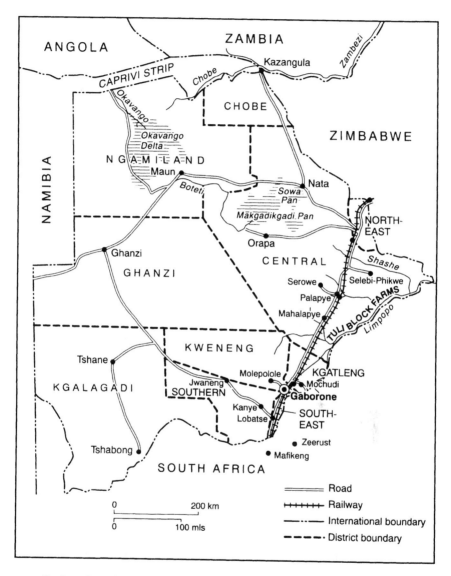

Soda ash and coal mining were the sectors of mining that were worst affected by price fluctuations. Nevertheless, coal production has increased, but remains relatively small as compared with that of South Africa and Zimbabwe. All the coal produced in Botswana is used for the production of power by the Botswana Power Corporation and for smelting for Bamangwate Concessions Limited (BCL). Prospects for exporting coal outside Botswana are insignificant given the low quality of the coal and the poor prevailing international prices (Silitshena and McLeod, 1998).

Despite the important role played by mining in Botswana, its linkages with the rest of the economy are limited owing to poor labour absorption capacity and its inability to use domestic intermediate inputs. Linkage with the rest of the economy has taken place mainly through the incomes that are generated. The national government is the main linkage between mining and the rest of the economy through its use of mineral revenues. This revenue accrues in the form of taxes, dividends and royalties. Through income created from mining the government has been able to develop new infrastructure and provide services, while ensuring stability in employment and income during periods of market volatility. The major private sector mining enterprise, Debswana, has however contributed to linkages by establishing polishing plants, setting up small-scale enterprises and supporting agriculture.

Despite the impressive contributions of mining to the economy, a number of issues are of major concern and need to be addressed. The mineral sector itself is not sufficiently diversified with substantial domination by diamond production. Some mines are not operating at a profit. Two mines – copper-nickel and soda ash – subsist only as a result of heavy government subsidies. Pollution caused by mining activities is a further major problem. The discharge of sulphur dioxide gases into the atmosphere from the processing activities of the Selebi-Phikwe copper mine, for example, has accounted for the disappearance of vegetation and contamination of soil in the surrounding area (Asare and Darkoh, 2001). Similar pollution and degradation of the environment is observable at Morupule where the burning of coal is exuding sulphur dioxide from a tall stack. A major aspect of policy is one of ensuring that environmental protection measures are internalised. i.e. paid for by operations themselves.

Botswana is highly dependent on mining for its export revenue, a situation which is not desirable. While in the short run the government may cushion the economy from volatile markets, in the long run Botswana must seek to diversify the economy away from minerals to other sectors that may include tourism and manufacturing. Overall, the challenge to Botswana is to use the mineral wealth wisely in order to promote a robust, sustainable and diversified economy.

Tourism

Most tourism activity in Botswana is nature-based and concentrated in the northern part of the country. The major attractions are national parks and reserves. The most important are the Moremi Wildlife Reserve and Chobe National Park. Visitors to these parks undertake game drives where they photograph the wild animals and birds. Game viewing and safari hunting are the major tourist activities. The world famous Okavango delta is a major tourist attraction with abundant wildlife and unique landscapes. It is a large swamp with an area of 16,000 square kilometres. Some of its unique features include a conglomeration of small islands with characteristic vegetation, which is home to a variety of wild animals. The people living in and around the delta are of different origins including the Baherero, Batawana, Bayei, Basarwa and Bakgalagadi. These people have unique cultures that are of interest to tourists. Overall, therefore, the Okavango delta is therefore an important tourism resource not only for Botswana but for promoting tourism in the entire southern African region.

A major concern in this sector is the extent to which a destination country benefits. In the case of tour packages foreign airlines and hotels receive the bulk of the receipts while destination countries on average receive only 22-25 per cent of the inclusive tour retail price paid by tourists. Overall, Botswana in 1998 generated 740,000 tourists which represents 3 per cent of the total for Africa as a whole. Botswana does not support mass tourism due to negative social and environmental impacts as have occurred in other countries (Bachman, 1988). Despite its potential for expansion of the tourism industry, Botswana has deliberately opted to adopt a low-volume, high-paying model of tourism development. This is aimed at reducing negative environmental impacts in order to conserve tourism resources.

The experience of several developing countries is that tourism development does not always achieve its desired goals. Tourism may be accompanied by local price increases as a result of free spending by tourists with limited benefits flowing to local communities. Mbaiwa (2000) has shown that in the case of tourism in the Okavango delta, foreign capital dominates the industry with very few benefits accruing to the local economy. Rather than promote conservation, tourism is engendering environmental problems. It is particularly feared that tourism may actually lead to cultural erosion as result of commodification of people's artefacts and practices. For the tourism sector to be sustainable, therefore, it is necessary carefully to monitor impacts on the environment, economy and society, so that the industry in Botswana does not erode the resources that support it, namely wildlife, culture and the environment.

Manufacturing

The manufacturing sector is crucial for the diversification of the Botswana economy and provision of employment opportunities. In other African countries agriculture absorbs much of the excess labour, but in Botswana, because of the steady decline in farming as an income generator and employment option, there is an emphasis on expanding manufacturing to absorb the growing excess labour. The manufacturing sector has been transformed over the last 25 years. At independence in 1966, it was dominated by the Botswana Meat Commission (BMC) which accounted for an estimated 95 per cent of the manufacturing value added. By 1979/80, the share of meat and meat products in total manufacturing value added had declined to about 31 per cent and by 1994 to about 15 per cent while the share of beverages, textiles and clothing had increased. The other sub-sectors which grew rapidly included metal products, tanning and leather working, and village industries such as beer brewing, baking, sewing and tool making. Village industries are important training grounds for formal sector entrepreneurs who will create further employment.

Manufacturing growth has been influenced by the overall growth of the economy which was accompanied by a rise in demand for manufactured goods, the introduction of the Financial Assistance Programme and import substitution. The sector's share of GDP has averaged 5.6 per cent between 1982/83 and 1998/89. Since the 1990s, however, the sector has experienced growth far below what had been expected, with an annual average of 4.2 per cent. This disappointing performance has been lower than that of non-mining output and overall GDP, which grew by 7.0 and 5.3 per cent per annum respectively. Employment in manufacturing decreased by 15 per cent from 26,300 to 22,100 between September 1991 and March 1993. In 1990, the sector accounted for 11.7 per cent of total formal employment but by 1995 this had decreased to 10 per cent.

A central factor influencing lower rates of growth is the slowdown of government spending, which had a knock-on effect for the construction sector signalling the end of Botswana's housing boom. Local manufacturing suffered further due to lower BMC throughput sales as the cattle off-take dropped due to the outbreak of cattle lung disease. Other factors within the region also had a negative impact on Botswana's manufacturing. The most significant factor related to macro-economic changes consequent upon Zimbabwe's Structural Adjustment Programme (SAP) which affected Botswana's exports to that country. The depreciation of the Zimbabwe dollar made Botswana's manufactured exports expensive in Zimbabwe dollar terms. Moreover, foreign exchange controls that had earlier constrained importation of inputs and encouraged relocation of

Zimbabwe firms in Botswana were liberalised. These economic reforms in Zimbabwe led to a rise in inflation, which cut real incomes and reduced that country's demand for Botswana's manufactured products. Also, the price of cotton lint that was sourced from Zimbabwe at half the world price before the SAP was lifted to world price, thereby raising domestic input costs and further reducing the international competitiveness of Botswana.

The spatial distribution of manufacturing in Botswana is concentrated in urban areas. Urban areas offer high incomes, larger markets, a better trained, educated workforce, good transport links and infrastructure that is relatively well developed. In 1999, the Central Statistics Office recorded 918 manufacturing establishments in operation. Of these, 474 (51.5 per cent) were located in Gaborone, 73 (8 per cent) in Francistown, 56 (6.1 per cent) in Selebi-Phikwe and 40 (4.4 per cent) in Lobatse. Thus, about 70 per cent of all manufacturing establishments in the country in 1999 are found in these four urban areas. Between 1985 and 1990, 475 manufacturing licences were issued according to the Central Statistics Office. Of these 31.6 per cent went to Gaborone, 6.7 per cent to Francistown, 8.2 per cent to Selebi-Phikwe and 15.8 per cent to Lobatse, which means that some 62.3 per cent of these licences went to these four towns. These figures indicate that government's rural industrial and growth centre policies, designed to promote development in all parts of the country to avoid an excessive concentration of industrial activity in a limited number of urban centres, are a long way from achieving their decentralised objectives.

Over several years, Botswana has provided development incentives for the industrial sector which have included, *inter alia*, the Financial Assistance Policy (FAP), the Selebi-Phikwe Regional Development Programme (SPRDP), the Citizen Reservation Policy, the Local Preference Scheme, now called the Local Procurement Programme (LLP), and tax incentives (Box 10.2). These support measures were deemed necessary in order to reduce dependence on mining. The initial policy frameworks supported intensive, import-substituting and export-oriented industries. A further objective was to create jobs for the unskilled and semi-skilled Batswana. Although some diversification took place, the industrial base remains narrow.

In 1998 Botswana produced its new industrial policy. This abandons the earlier emphasis on unskilled labour, and encourages high levels of skills and use of technology because of competitive pressure due to globalisation. There is an explicit shift away from producing for domestic to export markets. Emphasis and support in the future will be given to those industries that are viable and can compete in the international markets. Botswana will therefore continue to encourage foreign investments in the country in order to access management, technical expertise and export markets.

Box 10.2 Support for New Manufacturing Development in Botswana

A.   SELEBI-PHIKWE REGIONAL DEVELOPMENT PROJECT
      (SPRDP)

The Selebi-Phikwe Regional Development Project was introduced in 1988 to encourage investment in non-mining activity in the Selebi-Phikwe area. This was in order to promote regional diversification and sustainability beyond the life of the copper-nickel mine which supported the growth of the town. Financial assistance is extended to industrial projects locating in Selebi-Phikwe and meeting certain other conditions. However, with the reduction of company tax on manufacturing to 15 per cent, the financial incentives provided under the SPRDP have effectively been extended to the whole country.

B.   CITIZEN RESERVATION POLICY

The Citizen Reservation Policy was introduced in 1988 in order to promote active participation of citizens in industrial development. A number of industries that require relatively simple industrial skills and low levels of capital were reserved for citizens. These simple activities are the production of ordinary bread, cement and baked bricks, school uniforms, protective clothing and burglar bars, school furniture and the milling of sorghum. Commercial activities reserved for citizens include bottle stores, boutiques and liquor restaurants. The citizen reservation policy is still under review.

C.   THE LOCAL PREFERENCE SCHEME

The Local Preference Scheme (LPS) was introduced in 1976 and modified in 1986, in order to direct a substantial share of the purchases of Government, local authorities and parastatals to local manufacturers. The scheme is aimed at increasing production in Botswana and encouraging the use of local raw materials and labour in manufacturing. The firms which qualify for LPS are allowed a price advantage over foreign produced goods when tendering for government contracts. The price advantage is based on an assessment of the local content in production. The LPS was reviewed in 1994 to assess its impact and effectiveness. It was subsequently replaced by the Local Procurement Policy (LPP), effective from April 1997.

Source: Mpabanga, 1997.

The new policy framework was necessitated by the major changes that have taken place in South Africa and the potential the region has in terms of development and co-operation. The leaders of the SADC countries have passed resolutions to work closely together towards economic development of the area. The new WTO directives and the abandonment of sanctions against South Africa should provide international firms with freer access to the region creating competition for Botswana's products. Lastly, the past two decades have witnessed growth in the quality and quantity of goods and services, which requires that Botswana address international standards of efficiency and productivity in order to remain competitive.

Botswana's main non-mineral exports are beef, vehicles and textiles. The closure of the Hyundai motor-vehicle assembly plant in 2000 drastically reduced vehicle exports. The total value of exports between 1966-99 was P19.7m. In 1998 most of Botswana's exports were to UK and other European countries. Exports to African countries went mainly to South Africa and Zimbabwe. Botswana's export market is therefore characterised by a concentration on a few products and reliance on a few markets.

Botswana is dependent for trade on her neighbours as well as regional and bilateral agreements with the rest of the world. The regional agreements – the Southern African Customs Union (SACU) and the Southern African Development Community (SADC) – are discussed in chapter 15. The earlier SACU agreement reinforced industrial concentration in South Africa at the expense of other member countries and has therefore been under review (Hudson, 1981; Ochieng, 1981; chapter 15). Botswana will largely depend on SACU and SADC to gain export markets. It should be realised that these regional markets are likely to be highly competitive as tariffs decline due to World Trade Organisation (WTO) requirements. Overall, it is clear that Botswana will have to improve on her competitiveness if she is to retain her present market advantage.

Botswana has a Free Trade Agreement (FTA) with Zimbabwe and preferential access to the European Union (EU) under the Lomé Convention. While Botswana has benefited from the FTA with Zimbabwe, it has also experienced problems with the change of rules of origin, which were tightened in order to limit exports to Zimbabwe. The current economic and political crisis in Zimbabwe has seriously reduced its significance as an export market for Botswana. Botswana exports beef to the EU under the beef protocol of the Lomé Convention. It will continue to do so until 2007, when the trade aspect of Lomé IV expires; trade arrangements are due to be renegotiated in 2002.

In the short term Botswana's trade relations will depend on its existing arrangements and agreements but for the medium and long term it is evident that the country is already adjusting its economic structures in order to face the challenges that lie ahead. These adjustments are evident particularly in new industrial policy provisions and other macro-economic policies.

Challenges and Opportunities for Botswana within the SADC Region

Over the past decade the geopolitics of the southern African region have been transformed with the (re)admission of South Africa into the international community. Despite a host of regional problems, which include wars, HIV/AIDS and floods, the region has potential for growth because of its rich natural resources. At a recent meeting held in the Namibian capital Windhoek regional leaders agreed on strategies for strengthening economic integration. Within the SADC region, Botswana is poised to play a leading role because of its strong mineral-led economy and democratic society.

Botswana is, however, not immune to the impacts of the changing global and regional economic geography, especially the economic and political crisis in Zimbabwe. Fluctuations of commodity prices, foot and mouth disease in distant lands and in neighbouring countries, and the impact of HIV/AIDS on its small population and the economy constitute some key areas of concern for Botswana. In response to these and other challenges, Botswana is maintaining a strong economy supported by a stable government. Nevertheless, the country is looking beyond the minerals base in order to broaden the national base for economic and social development.

The Government of Botswana has been providing support in agriculture through infrastructural development, extension and marketing services, drought relief programmes and direct financial grants. This is justified because it is argued that the most needy people live in rural areas. In Botswana, according to the 1991 census, 54.3 per cent of the population was classed as rural. The population balance in Botswana is shifting, indicating that the poorest communities will not always be found in rural areas. Many of the funds dispersed to rural areas have not produced any significant results. Key questions that arise therefore are who should the target group be? When funds are allocated for agricultural development who benefits? It is evident that much of the money goes to wealthy rural farmers and not necessarily to the rural poor. Very few rural households own cattle and the same argument could be extended to the arable-farming sub-sector. Even when poor rural households take advantage of the funds of the Arable Lands Development Programme (ALDEP) and the Accelerated Rainfed Agriculture Programme (ARAP) they are often not successful. These area-based integrated agricultural programmes aimed at generating productive employment and raising rural incomes have had little impact in reducing poverty (Bank of Botswana, 1999).

In setting policies for the future, the Government of Botswana notes that a decline in agriculture should be seen as part of the normal process of development because this has occurred in the developed world. Through the diversification of its economy into tourism and manufacturing industry, it is hoped that more rapid rural development will take place and that problems of poverty and unemployment will be reduced. Several of these issues have formed the basis of discussion of the economic geography of Botswana in the foregoing sections.

Botswana cannot, however, expect to base its future economic growth upon domestic demand as occurred in the 1980s. Stronger economic growth in South Africa would benefit Botswana, although there is concern about South Africa's increasingly weak currency. The prospects for Zimbabwe are not good due to the economic and political crisis currently being experienced there. Looking to global markets, it is expected that exports to Europe and the USA will rise, albeit facing sharp competition as countries in the region implement World Trade Organisation provisions. Overall, therefore, the changing economic geography of Botswana is therefore going to be impacted both by trends in the SADC region and the rapidly globalising world economy.

References

African Business (2001), 'Around and About, Botswana Ranked as Best Emerging Market Risk', Issue 261, pp. 1-3

Asare B.K. and Darkoh M.B.K. (2001), 'Socio-economic and Environmental Impacts of Mining in Botswana: a Case Study of the Selibe-Phikwe Copper-Nickel Mine', *Eastern Africa Social Science Review* (in press).

Bachman, P. (1988), *Tourism in Kenya: A Basic Need for Whom?*, Peter Lang, Bern.

Bank of Botswana (1999), *Annual Review*, Government Printer, Gaborone.

De Blij, H.J. and Muller, P.O. (2000), *Geography: Realms and Concepts Regions 2000*, John Wiley and Sons, New York.

Gaolathe, N. (1997), 'Botswana's Boom and Recession Experience: a Discussion', in J.S. Salkin, D. Mpabanga, D. Cowan, J. Selwe, and M. Wright (eds), *Aspects of the Botswana Economy: Selected Papers,* James Currey, Oxford, pp. 37-52.

Hudson D.J. (1981), 'Botswana's Membership of the Southern African Customs Union, in C. Harvey (ed) *Papers on the Economy of Botswana, Studies in Economics of Africa,* Heinemann, London.

Kalabamu, F.T. (1998), 'Gender Neutral Housing Delivery Systems: a Critical Analysis of Housing in Botswana', in M. Mapetla, A. Larsson and A. Schlyter (eds), *Changing*

Gender Relations in Southern Africa: Issues of Urban Life, Institute of Southern African Studies, National University of Lesotho, pp. 56-76.

Mbaiwa, J.E. (2000), 'The Impact of Tourism in the Okavango Delta in the North-Western Botswana', unpublished paper presented at the Workshop on Climate Change, Biodiversity, Multi-species Production Systems and Sustainable Livelihoods in the Kalahari Region held at Maun, 9-13 October.

Ministry of Finance and Development Planning, *National Development Plans 5,6,7,8*, Government Printer, Gaborone.

Mpabanga, D. (1997) 'Constraints to Industrial Development', in J.S. Salkin, D. Mpabanga, D. Cowan, J. Selwe, and M. Wright (eds), *Aspects of the Botswana Economy: Selected Papers*, James Currey, Oxford, pp. 335-68.

Musyoki, A. (1998), 'Perceptions of Gender and Access to Housing in Botswana', in M. Mapetla, A. Larsson and A. Schlyter (eds), *Changing Gender Relations in Southern Africa: Issues of Urban Life*, Institute of Southern African Studies, National University of Lesotho, pp. 266-85.

Ochieng E.O. (1981), 'Botswana's Trade Structure Compared with those of other Small Countries', in C. Harvey (ed), *Papers on the Economy of Botswana, Studies in Economics of Africa*, Heinemann, London, pp. 115-30.

Parson, J. (1981), 'Cattle, Class and the State in Rural Botswana', *Journal of Southern African Studies*, vol. 7, pp. 236-55.

Pearce, F. (1993), 'Botswana Enclosing for Beef', *The Ecologist*, vol. 23 (1), pp. 25-9.

Silitshena, R.M.K. and McLeod, G. (1998), *Botswana: a Physical, Social and Economic Geography*, Longman, Gaborone.

11 Lesotho: Peripheral Dependence, Poverty and Political Instability

ANTHONY LEMON

Introduction

Since independence in 1966, Lesotho's peripheral position in the regional space-economy has been comparable with that of the larger South African bantustans such as Transkei and KwaZulu in the apartheid era. Four decades of conflict with the Boers of the Orange Free State in the mid-nineteenth century left Basutoland, as it was known in colonial times, shorn of the fertile cornlands west of the Caledon River. Its boundaries have remained unchanged since 1869, and its current overpopulation is directly related to the loss of the 'conquered territories'. An influx of refugees from the Anglo-Boer wars compounded the problem, contributing to the establishment of a pattern of labour migration in the early twentieth century which has persisted to this day. The migrant labour system in turn contributed to underdevelopment, turning Basutoland from an exporter of grain to a labour reserve (Murray, 1981).

British colonial neglect is often attributed to the assumption that Basutoland, along with the other High Commission territories of Bechuanaland and Swaziland, would eventually be incorporated in South Africa (Bardill and Cobbe, 1985, pp. 244-5). Long before the nature of apartheid became clear, however, the threat of incorporation into what was considered to be an unjust and hostile South Africa led to the growth of national identity in Basutoland (Rosenberg, 2001). British misgivings about the 'viability' of small colonial territories as sovereign states were perhaps nowhere more understandable than in Lesotho. This small, largely mountainous country, landlocked and entirely surrounded by South Africa, has a very weak productive base. Its major natural resources are water and abundant labour, and it currently depends on four major sources of capital formation: migrant remittances, dividends from the Southern African

Customs Union (SACU), foreign aid and export of water through the Lesotho Highlands Water Project (LHWP).

Population has been growing by 2.2 per cent annually in the 1990s, and reached about 2.2m. in 2001. The labour force is, however, growing by 2.6 per cent annually. For most of the past two decades the economy has grown rapidly, albeit from a very small base: real GDP grew by an average of 3.7 per cent p.a. in the decade 1979-89 and 5.9 per cent p.a. between 1989 and 1997. Manufacturing and services fuelled growth in the late 1980s and early 1990s, whereas in the mid-1990s it was driven by the construction of the LHWP. A sharp economic reversal followed in 1998, when GDP declined by 5.0 per cent owing to civil unrest following the 1998 election which led to intervention by troops from South Africa and Botswana and widespread destruction of property in the capital, Maseru. Growth resumed more slowly in 1999 (2.5 per cent) and the World Bank (2000a) predicts an average 4.1 per cent growth rate in 1999-2003.

In Lesotho, however, the GNP has traditionally been more than twice as large as the GDP owing largely to the importance of remittances from Basotho migrants. For reasons explained below, this has meant that the GNP has grown much more slowly that the GDP in recent years. In per capita terms, GNP grew by an average of just over 1.4 per cent p.a. in 1979-1997, but dropped by 11.1 per cent in 1998 and 1.5 per cent in 1999. The World Bank (2000a) forecasts a much healthier 3.1 per cent p.a. growth in 1999-2003. This somewhat optimistic prediction is evidently influenced by the importance of construction work on the LHWP which has helped the GDP to climb to 80 per cent of GNP in recent years.

Lesotho has a limited formal employment sector and unemployment is estimated to be over 40 per cent. Poverty is widespread, with almost half the population living below the national poverty line. There is considerable inequality, with 45 per cent of total income accruing to 10 per cent of the people and a Gini index of 0.57 (Lesotho, 2000). Some of the worst rural poverty is found in remote highland districts and the Senqu valley, but as people unable to survive on the land are forced into the towns there is an inevitable urbanisation of poverty. Pockets of extreme poverty and marginalisation are beginning to emerge in Maseru and smaller centres such as Mohale's Hook (UNDP, 1997). Maseru's Sechaba Consultants estimated in 1993 that 28 per cent of households in Maseru were below the poverty line (Romaya and Brown, 1999). In some peri-urban and urban areas the traditional norms of Sotho culture that have hitherto enabled the poor to survive through access to land for subsistence and mutual assistance for food and credit are no longer upheld. Urban poverty is also compounded by unemployed returning migrant workers, known as *Mekoao*, who often linger alone in urban areas, reluctant to return home to rural communities empty-handed. Overall, the rural population is declining in absolute terms while the population in the towns, which is growing in both number and size, has been inceasing by 6.3 per cent p.a. since 1986, leading to a functional urbanisation level of about 30 per cent by 1996 (Marais, 1999).

The Rural Economy

Mountainous terrain, a thin soil layer and limited vegetative cover makes for a fragile ecology in much of Lesotho. Pressures on land from human and animal populations have led to serious environmental problems. These include loss of topsoil which is eroding agricultural productivity and increasing siltation in the Caledon River, increased gully erosion, loss of tree cover owing to excessive cutting for firewood and damage to saplings from animals, and overgrazing. Four out of five rural households own livestock, mainly cattle, sheep and goats. Many also have a horse, two or more donkeys and chickens. The considerable potential of the livestock sector has been little exploited, but there is some export of cattle, wool and mohair. The opening of an abattoir in 1983 was intended, having satisfied domestic demand, to encourage expansion of production for regional and European Union demand, following the successful example of Botswana (chapter 10), but progress has been limited.

Average yields of maize and sorghum, the major crops, have declined to half their 1970 levels, and crop production could cease altogether if steps are not taken to reverse the decline in soil fertility (FAO, 2001). Despite fertiliser subsidies, prices are out of reach for many rural households. The same is true of seeds, which are also subsidised: most of the maize varieties used are local or recycled hybrid seed and almost no improved sorghum seed is used. Mochebelele and Winter-Nelson (2000) found that technical efficiency appears to be greater in households benefiting from migrant worker remittances, despite loss of male labour.

Although only about 11 per cent of Lesotho can support arable cultivation, a further 66 per cent is suitable for pasture, and subsistence farming remains the primary occupation for most Basotho. However the contribution of agriculture to GDP has diminished from 47 per cent in 1970 to only 17 per cent in 2000. Its share of total exports is even smaller - 13 per cent in 1999 - and far exceeded by agricultural imports which supplied half Lesotho's cereal requirements in the years 1995-2000. The neighbouring Free State Province of South Africa, which includes the 'conquered territories' referred to above, produces more maize than any other province in South Africa and is the natural supplier to Lesotho. Poor weather conditions in 2000-01 reduced cereal production by 60 per cent, leaving some 10-15 per cent of rural households in the worst affected districts - Mokhotlong, Thaba-Tseka, Mohale's Hoek and Quthing (Figure 11.1) - in need of food and seed assistance in 2001/2 (FAO, 2001). Elsewhere most households, though poor, have adequate survival strategies including sale of livestock, remittances from family members working abroad or in urban areas, local wage labour and informal sector activity.

Figure 11.1 Lesotho

To counter the problem of recurrent droughts, which hit output severely in 1990-93 and again in 1998, the government is implementing a food security programme based on the development of small-scale irrigated agricultural schemes and improved rural water supplies. The Lesotho Agricultural Development Bank, established in 1980 as the sole source of farmers' credit, was intended to play an important part in the self-sufficiency programme, but it was declared bankrupt and closed in 1998. Alternative sources of credit are clearly essential.

Another pressing need is land tenure reform, which is currently under review. The 1979 Land Act sought to improve productivity by conferring greater security of tenure, but few farmers registered their land for tenure upgrade as required (Kishindu, 1994). Kishindu argues for greater priority to be given to the control of soil erosion and for consolidation of land holdings into more viable units, but the latter poses obvious problems given the lack of alternative employment. The Land Reform Commission proposes abolishing customary tenure under which the chiefs control land, and making all land leasehold. Related problems include the exclusion of women from property rights which affects not only ownership and inheritance but also ability to raise credit. Less educated farmers are also precluded from recourse to their current rights because legislation governing land is in English.

Mines and Migrants

Lesotho itself has few mineral resources. Its clay deposits are being exploited for the manufacture of bricks, high-quality ceramic ware and tiles. The only mineral export is diamonds from the mine at Letsang-la-Terae, opened in 1977 by De Beers. By 1980 diamonds accounted for 55 per cent of Lesotho's limited exports, but falling world prices led to the closure of the mine in 1983. It was reopened in 1999 under new ownership, with a 24 per cent government holding. In 2000 the government was reported to be considering proposals from two Canadian companies to open further diamond mines at Kao and Liqhobong (EIU, 2000). Co-operatives using labour-intensive methods still recover a small quantity of low-grade diamonds.

Although Lesotho still provides a quarter of the South African mine labour force, and remains the main foreign supplier of workers, the number of Basotho gaining employment has declined from 103,700 in 1995 to 68,600 in 1999 (IMF, 2001a) and 60,000 in 2000 (EIU, 2001). This decrease reflects low gold prices which result in mine closures in South Africa and increased reliance on South Africa's own workforce. Miners' remittances have fallen from 60 per cent of GDP in the late 1980s to 26 per cent in 1999, and are predicted to fall in absolute value by almost a quarter between 1999 and 2003 (IMF, 2001b). In 1990, in response to pressure, the Lesotho government altered the terms of its unpopular Deferred Payment Scheme, halving the percentage of migrant workers' wages which had to be placed in a special account at Lesotho Bank from 60 to 30 per cent. The Lesotho government has lost further revenue since the end of 1995 when South Africa announced a policy of granting permanent residency rights to migrant workers who had voted in the 1994 election and were in the country before 30 June 1996.

Those still able to obtain employment in the South African mines have also benefited from increases in historically low wages. They grew rapidly in the mid-1970s, as the Chamber of Mines sought to recruit more workers from South Africa itself, where the mines had to compete with wages in manufacturing and services, and more steadily thereafter. However the rapid growth of sub-contracting in South African mines since 1990 has led to lower average wages, loss of benefits such as medical aid, sick leave and compensation for injury and/or death, and deteriorating conditions of work. It has also undermined the National Union of Miners (Crush et al., 2001).

Despite declining mine employment the overall number of Basotho migrant workers in South Africa is still increasing and remittances remain the most common coping mechanism for rural households (FAO, 2001). The decline in mine remittances has been accompanied by an increase in the number of female migrants who find domestic work, farm, factory and sales jobs. This reflects a degree of household role reversal: in the case of married Basotho farm workers, seventy per cent had an unemployed spouse (Uliki and Crush, 2000).

The Lesotho Highlands Water Project

This long-planned project to divert water from Lesotho's rivers to South Africa was agreed in 1986 and reconfirmed in 1996 after a change of government in Lesotho. Phase 1A includes the Katse Dam, the highest in Africa, which was officially opened in 1998, with a water-flow capacity of 18 cubic metres per second, and a smaller dam and hydro-electric power plant at Muela which has made Lesotho self-sufficient in electricity. Export is possible, but costs are currently higher than in South Africa. Completion of phase 1B, which will increase water-flow to 30 cubic metres per second, is scheduled for 2003. South Africa pays monthly royalties plus a unit cost component based on the volume of water delivered; revenue may eventually be as much as one-third of Lesotho's GDP (Waites, 2000). Much of the funding for both phases has come from South African capital and from the water users themselves. Lesotho's share has been tiny, only $25m. of a projected $1,100m. in the case of phase 1B. Negotiations have started over a second phase of the project. If all five phases are completed by 2020 as envisaged, water-flow capacity would rise to 70 cubic metres per second and total electricity output from the current 80 mw. to 200 mw. far beyond Lesotho's domestic needs.

Debates surrounding the transformation wrought by this massive project - one of the largest civil engineering projects in the world - have been characterised by both quantitative and qualitative approaches (Matlosa, 1998a). The Lesotho government and the World Bank stress the benefits accruing to Lesotho, including export revenue, electricity self-sufficiency, improved rural infrastructure in the project area and, to date, some 7,000 jobs. More qualitative assessments advanced by international and local NGOs have focussed on socio-economic issues, including the stresses of forced resettlement and dislocation (Thabane, 2000; Waites, 2000; Nel and Illgner, 2001).

The Manufacturing Sector

Prior to 1980 manufacturing activity was minimal, but clothing and textiles alone accounted for 29,000 jobs in March 2001 (Mosisili, 2001) and manufacturing contributes more than three-quarters of exports in most years, a much greater proportion than in any other country in southern Africa. Lesotho has abundant labour with one of the lowest adult illiteracy rates in Africa (17 per cent). Duty-free access to the South African and European Union markets is important, although South Africa actively discouraged the development of competing industries in Lesotho until the mid-1990s. Foreign companies have benefited from generous allowances and tax 'holidays', the provision of industrial infrastructure and the construction of industrial estates in Maseru and Maputsoe, with others planned. With the widespread imposition of sanctions against South Africa from the late 1980s, Lesotho intensified its efforts to encourage South African firms to relocate, announcing incentives in 1996.

By 1989 the Lesotho National Development Corporation was supporting 51 companies employing more than 10,000 workers in a wide variety of small industries including textile and footwear companies in the Thetsane industrial area in southern Maseru which has considerable land available for future expansion. The Basotho Enterprises Development Corporation has also provided finance to local entrepreneurs, with Canadian support, creating up to 3,000 jobs between 1975 and 1995. The World Bank's private-sector institution, the International Finance Corporation, is also active in Lesotho, seeking to develop small and medium-sized enterprises, to revive extractive industries and to put existing industrial assets to more productive use.

The clothing and textile sector has grown particularly fast, and Lesotho is now the largest African supplier of clothing to the USA. From April 2001 the textile sector should be the main beneficiary of the Africa Growth and Opportunities Act which allows duty-free access to the USA for certain products. It has already benefited from relocation by Asian manufacturers to avoid developed country quota restrictions. In 2000 the economy received a major boost with the announcement of a $100m. investment by a Taiwanese company, Nien Hsing, in a giant denim mill in Maseru which should create 5,000 jobs by the time it is fully operational in 2004. The scale of this investment may encourage others to follow.

Lesotho's industrialisation strategy is not unproblematic. It continues to rest on cheap female labour, reflecting social convention and the failure of successive agricultural projects as well as declining household income with the decrease in mine recruitment (Baylies and Wright, 1993). The external environment is another source of risk, given the extreme openness of Lesotho's economy. Export-based industry would be seriously affected by the rise of protectionist trends in its main markets, although this danger currently seems remote. Many export-based firms could quickly and easily relocate to other places as the business environment changes. Political unrest in September 1998 caused substantial damage to businesses, severely denting investor confidence and resulting in the loss of 4,000 jobs. Recovery has been rapid, but this incident demonstrated underlying political fragility. The strong business environment needed to retain and attract investors demands political stability which has hitherto proved elusive.

Industrial exports have helped to diversify Lesotho's export markets, with the USA beginning to erode South Africa's two-thirds share. Imports remain overwhelmingly South African, but Asia accounted for 13 per cent of imports in 1999, reflecting imports by Asian firms of raw materials from their home countries, especially Taiwan. Trade with the European Union amounted to less than one per cent in both directions. The ratio between imports and exports is predicted to improve from 4.5 in 1999 to 3.1 in 2002 (EIU, 2001), and may do slightly better than this if current manufacturing growth continues. Declining remittances and now declining revenue receipts from SACU (chapter 15) imply a continuing current account deficit of around 20 per cent of GNP. Revenues from the LHWP are not expected to exceed the costs of imports associated with the project until 2002/3, but should thereafter help to reduce the deficit. Development aid has fallen substantially since the early 1990s as more assistance has been directed to post-apartheid South Africa (Love, 1999).

Health and Education

Lesotho has not been immune to the regional HIV/AIDS crisis, and the migrant labour system has undoubtedly contributed to it. Almost a quarter of adults aged 15-49 years were HIV positive in 1999, and infection rates are still rising. By the end of 1999 at least 16,000 people had died of AIDS, and average life expectancy had been reduced from 69 years in 1991 to 52. Anticipated mortality from AIDS implies that Lesotho will begin to lose about 2 per cent of its working-age population each year by 2004, and the annual loss of economic growth is projected to rise from 0.6 per cent in 2001 to 2.7 per cent in 2015 (World Bank, 2000b). With the public sector accounting for 43 per cent of all employment, HIV/AIDS-related absenteeism, medical treatment, pension payments and staff replacement and training will increase the government wage bill. The impact of AIDS will also be harder on the poor, who are less educated about prevention and more vulnerable to rising medical costs and loss of income. Lesotho lacks both the resources to provide anti-retroviral drugs and the means to distribute them effectively through the country's rudimentary health system. It has responded with public education campaigns and, with donor support, the distribution of prophylactics to individuals in high-risk groups.

Otherwise the level of health is relatively good, and the climate does not encourage tropical diseases. Waterborne diseases are spread by poor sanitation and there is malnutrition, which is exacerbated in drought years. The government is committed to increasing spending on health in real terms, and primary health care in rural areas is improving. Currently some 80 per cent of the population enjoy access to health services, compared with 57 per cent in sub-Saharan Africa generally. In June 2000 the World Bank approved assistance for a nine-year health reform programme.

Good basic education is important if Lesotho is to maintain its competitiveness in the manufacturing sector. Whilst the country undoubtedly needs more skilled people too, increasing concentration of development assistance on human resource investment may, given the attractions of Lesotho's larger neighbour, lead to out-migration and benefit South African expansion (Love, 1999). The World Bank is supporting a twelve-year programme to improve educational facilities and quality, and this may help to reverse declining examination performance in the secondary sector. Provision of vocational and technical post-secondary education is improving, but still inadequate. Enrolment in primary education (69 per cent) is high for sub-Saharan Africa, thanks largely to the churches, but male enrolment has traditionally been lower than female, with most herdboys excluded from the classroom. Falling enrolment in recent years was reversed in 2000 with the introduction of free primary education for new first-year pupils, a policy which will be extended to cover all primary classes over the next six years. The additional resources provided by the government, UNICEF and the World Food Programme have so far been inadequate to cope with the demand, and the pupil-teacher ratio is currently around 70:1 (EIU, 2000).

Economic Policies, National Identity and Political Stability

The Lesotho government has gained IMF approval for its programmes of privatisation and structural reforms, and was granted a three-year Poverty Reduction Growth Facility in 2001. The government has agreed to pursue a path of fiscal consolidation through increases in tax revenue, cuts in public expenditure and public sector reforms. It is seeking to improve the domestic revenue base, and to anticipate declining customs revenues from SACU, by introducing value-added tax. Privatisation has been hindered by lack of demand rather than bureaucracy, but it has proceeded faster than in other countries in southern Africa, with support from the World Bank. More than fifty state-owned enterprises and parastatals have been identified for divestiture or closure. Completed privatisations include telecommunications and two state banks, and attention is now focused on electricity and water companies.

Such structural reforms do little to address immediate problems of poverty and social exclusion. An independent report by Sechaba Consultants (2000) concluded that despite rapid economic growth during much of the 1990s, the poor had benefited little. Poverty in highland areas may have increased despite improved access to services, owing to rising unemployment and declining agricultural conditions. Unemployment and poverty contribute to the fragility of Lesotho's democracy, which has been the subject of several recent academic analyses. Quinlan (1996) argues that Sotho nationalism cannot be sustained in the face of the incapacity of the state to improve the existence of its citizens, and Coplan and Quinlan (1997) argue that the sovereign state may not be the most ideal or beneficial political embodiment of a national identity. Matlosa (1998b) argues that the very process of political liberalisation unleashes demands and raises expectations which the state is unable to meet, leading to increasing conflict between the state and organs of civil society.

Under conditions of economic scarcity, the state inevitably becomes both the channel and the source of accumulation, and competition for the control of the state becomes keen (Akokpari, 1998). This certainly accords with recent experience. Following unrest after the 1998 elections, an interim political authority was established to recommend electoral reform, but the government wants to reduce the proposed number of members of parliament to be elected by proportional representation and objects to the proposed use of fingerprint technology in voter registration. The postponement of elections led to a cooling in Lesotho's relations with both SADC and European donors during 2001 (EIU, 2001).

Lesotho's record of political instability since independence has led some to raise the issue of incorporation into South Africa, questioning the value of sovereignty and the meaning of the Free State border (Lemon, 1996; Makoa, 1996; Coplan, 2001). Influential local consultants argue that 'in the long run it is difficult to imagine a more effective poverty reduction strategy than integration' (Sechaba Consultants, 2000, p. 194) which they see as bringing better pensions, health care, education and a larger job market. Sovereignty is not without its benefits for small states (Lemon, 1993), and there is arguably a genuine sense of nationhood among the Basotho people, but this is certainly a debate which needs to become part of Lesotho's national agenda in post-apartheid southern Africa.

Conclusion

As a sovereign state Lesotho is one of the strangest products of the partition of Africa. Its colonial inheritance was unpromising: a largely mountainous territory shorn of much of its most fertile lowlands and a landlocked situation surrounded by a single powerful neighbour on which it was dependent for trade and employment. Its relatively confrontational stance towards apartheid South Africa until the mid-1980s helped the government to attract foreign aid despite its lack of legitimacy; it also suggested that political independence is not wholly negated by economic dependence. The development of manufacturing since 1980 and of the Lesotho Highlands Water Project since 1986 have begun to transform an economy which hitherto depended largely on migrant labour remittances and revenues from SACU, both of which are now in steep decline. Tourism, for which considerable potential exists in this 'Switzerland of Africa', remains largely undeveloped.

Lesotho's very existence as a sovereign state makes less sense in a post-apartheid southern Africa, and the instability of its restored democracy can only serve to undermine its economic advancement. The vested interests of its ruling classes and the national identity of its people may guarantee continuing statehood, while South Africa may be less than eager to acquire responsibility for more than two million more poor people. Yet Lesotho is in something of a catch-22 situation: unemployment and poverty contribute to its political instability, which in turn threatens the more positive economic achievements of the past two decades. A combination of effective domestic political leadership and firm regional and international support is needed if the country is to build on those achievements, improve the prospects of its people and in doing so sustain the nationalism which underpins Lesotho's continuing statehood.

References

Akokpari, J.K. (1998), 'A Theoretical Perspective on Prospects for Democratic Stability in Lesotho', *Lesotho Social Science Review*, vol. 4, pp. 65-82.

Bardill, J.E. and Cobbe, J.H. (1985), *Lesotho: Dilemmas of Dependence in Southern Africa*, Westview, Boulder, Colorado.

Baylies, C. and Wright, C. (1993), 'Female Labour in the Textile and Clothing Industry of Lesotho', *African Affairs*, vol. 92, pp. 577-91.

Coplan, D. (2001), 'A River runs Through it: the Meaning of the Lesotho-Free State Border', *African Affairs*, vol. 100, pp. 81-116.

Coplan, D. and Quinlan, T. (1997), 'A Chief by the People: Nation versus State in Lesotho', *Africa*, vol. 67, pp. 27-60.

Crush, J., Ulicki, T., Tseane, T. and Jansen van Veuren, E. (2001), 'Undermining Labour: the Rise of Sub-contracting in South African Gold Mines', *Journal of Southern African Studies*, vol. 27, pp. 5-31.

EIU (2000), *Country Profile 2000: Lesotho*, Economist Intelligence Unit, London.

EIU (2001), *Country Report: Lesotho*, Economist Intelligence Unit, London.

FAO (2001), FAO/WFP Crop and Food Supply Assessment Mission to Lesotho: Press release, accessed at: http://allafrica.com/stories/printable/200106210060.html.

IMF (2001a), *Lesotho: Statistical Annex*, Country Report 01/80, June 2001.

IMF (2001b), *Lesotho: 2000 Article IV Consultation and Request for a Three-Year Arrangement under the Poverty Reduction and Growth Facility. Staff Report; Staff Statement; Public Information Notice*, IMF Country Report 01/79, June 2001.

Kishindo, P. (1994), 'Land Reform and Agricultural Development', *Journal for Rural Development*, vol. 13, pp. 319-26.

Lemon, A. (1993), 'The Political and Strategic Vulnerability of Small Island States', in D.G. Lockhart, D. Drakakis Smith, and J. Schembri, J. (eds), *The Development Process in Small Island States*, Routledge, London, pp. 38-56.

Lemon, A. (1996), 'Lesotho and the new South Africa: the Question of Incorporation', *Geographical Journal*, vol. 162, pp. 263-72.

Lesotho (2000), *Interim Poverty Reduction Strategy*, Ministry of Development Planning, Maseru.

Love, R. (1999), 'Changing Aid Patterns in Southern Africa', *Development in Practice*, vol. 9, pp. 296-309.

Makoa, F.K. (1996), 'Debates about Lesotho's Incorporation into the Republic of South Africa: Ideology versus National Survival', *Africa Insight*, vol. 26, pp. 347-53.

Marais, L. (1999), 'Urbanisation Dynamics in Lesotho', *Africa Insight*, vol. 29, pp. 11-16.

Matlosa, K. (1998a), 'Changing Socio-economic Setting of the Highlands Regions as a Result of the Lesotho Highlands Water Project', *Transformation*, vol. 37, pp. 29-45.

Matlosa, K. (1998b), 'Democracy and Conflict in Post-apartheid Southern Africa: Dilemmas of Social Change in Small States', *International Affairs*, vol. 74, pp. 319-37.

Mochebelele, M.T. and Winter-Nelson, A. (2000), 'Migrant Labour and Farm Technical Efficiency in Lesotho', *World Development*, vol. 28, pp. 143-53.

Mosisili, P. (2001), Address reported in *Mpheme/The Survivor* (Maseru), 27 June 2001.

Murray, C. (1981), *Families Divided: the Impact of Labour Migration on Lesotho,* Cambridge University Press, Cambridge.

Nel, E. and Illgner, P. (2001), 'Tapping Lesotho's "White Gold": Inter-basin Water Transfer in Southern Africa', *Geography*, vol. 86, pp. 163-67.

Quinlan, T. (1996), 'The State and National Identity in Lesotho', *The Journal of Legal Pluralism and Unofficial Law*, vol. 37/38, pp. 377-405.

Romaya, S. and Brown, A. (1999), 'City Profile: Maseru', *Cities*, vol. 16, pp. 123-33.

Rosenberg, S. (2001), '"The Justice of Queen Victoria": Boer Oppression and the Emergence of National Identity in Lesotho', *National Identities*, vol. 3, pp. 133-53.

Sechaba Consultants (2000), *Poverty and Livelihoods in Lesotho, 2000: more than a Mapping Exercise*, Maseru.

Thabane, M. (2000), 'Shifts from Old to New Social and Ecological Environments in the Lesotho Highlands Water Scheme: Relocating Residents of the Mohale Dam Area', *Journal of Southern African Studies*, vol. 26, pp. 633-54.

Uliki, T. and Crush, J. (2000), 'Gender, Farmwork and Women's Migration from Lesotho to the new South Africa', *Canadian Journal of African Studies*, vol. 34, pp. 64-79.

UNDP (1997), *Assessment of Urban Poverty*, UNDP, Maseru.

Waites, B. (2000), 'The Lesotho Highlands Water Project', *Geography*, vol. 85, pp. 369-74.

World Bank (2000a), *Lesotho at a Glance*, Washington DC.

World Bank (2000b), *Lesotho: the Development Impact of HIV/AIDS*, Washington DC.

12 Swaziland: Changing Economic Geography

MIRANDA MILES-MAFAFO

Introduction

In common with many other African countries, Swaziland has experienced shifts in its economic geography that can be attributed to a wide range of external and internal socio-political and economic factors. The country has enjoyed a period of thirty-five years of independent rule. Globalisation presents this tiny landlocked kingdom with many critical challenges. These challenges not only have significant impacts for the economic geography of Swaziland, but also for her position in the competitive global market. The task in this chapter is to present an overview of Swaziland's changing economic geography and of new visions and potential for reviving economic growth, reducing inequalities and reducing poverty.

Background History and Politics

Swaziland is a small landlocked country in southern Africa, with an area of 17,364 km². Surrounded mainly by the Republic of South Africa, and by Mozambique in the north-east, Swaziland possesses a remarkably wide range of physical environments (Figure 12.1) which provide the basis for different forms of development initiatives. Overall, it has been argued that three interlocking factors provide the basis for Swaziland's political history: the emergence of the Swazi as an independent nation, the gradual restriction of the territory under Swazi rule, and the confrontation between the Swazi and the whites together with their subsequent reaction to white hegemony (Funnell, 1991). Detailed accounts of the consolidation of Swazi traditional power in the early nineteenth century are offered by Bonner (1983) and Booth (1983). Importantly, by the 1880s, a series of conventions established primarily by the British and the Boers demarcated the land area ruled by the Swazi and the country's current boundaries. Indeed, as a consequence of the geopolitics of the region, first the Boers in 1890 and then the British in 1902 annexed the country, the latter making it a High Commission territory by 1906.

Figure 12.1 Swaziland

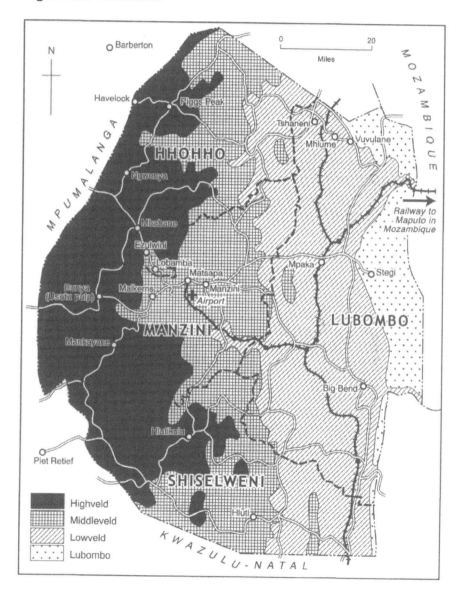

The key mission of the late Swazi King Sobhuza II became the achievement of independence from the British and the regaining of their land for the Swazi people. Independence came in 1968 and ushered in a period of growing economic dependence on South Africa in addition to growing political ambivalence, particularly in the past two decades under the rule of the reigning King Mswati III. It is evident that the period 1982-86 between the death of Sobhuza II and the crowning of Mswati III was one clearly marked by closer liaisons with South Africa both at a political and economic level. Swaziland's geographical position has placed the country in a precarious position with respect to other southern African countries. Membership of the Southern African Customs Union (SACU), which is dominated by South Africa, together with the country's proximity to South Africa and the large trade flows between the two countries, means that the economy of Swaziland is greatly affected by changes that occur in South Africa. Similarly, political tensions and reforms in South Africa have had significant impacts on Swaziland's political and economic stability. After South Africa's transition to democracy in 1994, the kingdom was hit by a spate of strikes and political violence associated with calls for a multi-party system and a constitution that would restrict Swaziland's monarchy to a symbolic role in government.

Economic Development and Change

The decade following independence in 1968 was one in which Swaziland experienced a high rate of economic growth. The period 1968-86 was characterised by remarkable rates of growth and important structural changes in the Swazi economy, as well as diverse changes in the country's political and social fabric. Many of these changes were related to South Africa's controversial position within the global economy and the imposition of economic sanctions by various states and groups. Characteristic of the 1968-86 period was a complete transference from dependence on British capital to dependence on South African capital (Atkinson, 1992). The escalation of economic sanctions on the Republic of South Africa in 1986 formed a watershed for the economic boom of post-colonial Swaziland.

The kingdom's economic life is marked by a dualism between a modern sector geared for export production and the subsistence sector. As compared with many other African countries, Swaziland's economy is relatively developed and diversified, but it remains heavily dependent on agriculture, especially the capitalist estate sector. It is an economy that is substantially vulnerable to fluctuations in the social and economic development of South Africa. Overall, Swaziland receives up to 80 per cent of her imports from

South Africa, and sends up to 70 per cent of her exports to South Africa, either for use there or for re-export (Esterhuysen, 1998). Furthermore, South Africa and Swaziland co-operate closely with respect to physical and bulk infrastructure such as rail, telecommunications, electricity and water.

Atkinson (1992) points out that while Swaziland has some of the attractions of a 'small is beautiful' economy, the country is extremely vulnerable. Its economic openness, marked by high dependence upon agricultural exports, is compounded by the geography of trade. The large proportion of imports that passes though South Africa depends upon the use of South African ports. The Swazi economy is also vulnerable to global and regional variations in economic activity which rapidly diffuse into Swaziland with retrenchments and other common symptoms of economic stress (Funnell, 1991). An assessment of the changing economic geography of Swaziland thus requires recognition of the fact that Swaziland is part of a global community and of the need to assess the impact of globalisation on the national economy (Madonsela and Dlamini, 2001).

In his 2001 budget speech, Swaziland's finance minister, Dr. Sibusiso Dlamini admitted that economic growth had declined in 2000 as compared with 1999 (Government of Swaziland, 2001). Among other things this decline has been attributed to the poor performance of some major industries and low agricultural production on Swazi Nation Land (Madonsela and Dlamini, 2001). Although in his budget speech the minister refers to the 1.2 per cent decline in economic growth rate as 'slight', Madonsela and Dlamini (2001) argue that this is a gross understatement of the country's economic difficulties. They submit that a decline of this magnitude has very grave implications for incomes, nutritional levels of the population and for poverty.

Agriculture

Agriculture is the backbone of Swaziland's economy and the main source of growth and employment. Evidence suggests that up to 11 per cent of the population are directly dependent on the sugar industry alone, which is the country's single largest employer and supplier of the leading export commodity (Esterhuysen, 1998). One reason why sugar is a resilient export crop is because of its ability to withstand drought conditions. Furthermore, there has been an increase in sugar cane production due to increased output from small-scale producers. Although statistics would suggest that the contribution of agriculture to GDP has declined relatively, this is in line with the fifth national development plan that expected the manufacturing sector to supersede agriculture. Yet it must be understood that Swaziland's manufacturing sector is little more than an extension of agriculture.

Overall, it is dominated by processing activities such as sugar milling, woodpulp processing and fruit canning (Funnell, 1991).

Other important agricultural products include cotton, maize, tobacco, rice, vegetables, citrus fruits and pineapples. In addition, Swaziland has Africa's largest man-made forest in the highveld region. This earns a substantial share of total export revenue through the sale of wood and wood pulp.

Mining

Prior to 1940 little social and economic development occurred in Swaziland (Fair, 1981). Until then, mining was virtually the only industry in Swaziland. This involved the exploitation of minerals such as tin, gold and asbestos whose production levels were generally low. Mining remains significant, but the contribution of the mining sector to overall production has declined despite the country's rich mineral endowment. This is due mainly to the closure of the Ngwenya iron ore mine in 1978 and the declining production of the Havelock asbestos mine since the mid-1970s. The last iron-ore shipments were made in 1980, and the closure of the mine resulted in the decline and decay of the rail link to Maputo in Mozambique. The major focus of Swazi mining has now shifted towards coal and diamonds. After long and difficult negotiations, diamond mining began in 1984 at the Dvokolwako diamond mine which has has become a key growth sector of the economy. Considerable potential also exists for the Mpaka coal mine, which is said to have anthracite reserves of 1 bn. tonnes. However the exploitation of these minerals is very dependent upon international market conditions. Currently all mining rights are vested in the king and royalties pass to the Tisuka takaNgwane Fund, an organization established in 1975 to control funds from mineral royalties. Many of the mining concessions are in the hands of international or South African companies.

Manufacturing

Until the late 1980s, approximately 80 per cent of the value added by the manufacturing sector was derived from industries processing Swaziland's agricultural and forestry products, notably wood pulp, timber, sugar, fruit, cotton and meat. By 2000 the country produced a wide range of goods, mainly for South African markets, including sweets, beverages, textiles, footwear, beer products, knitwear, safety glass and bricks. The majority of this activity takes place at the Matsapha industrial estate (Figure 12.1). The importance of manufacturing is demonstrated by the fact that it contributes about half of Swaziland's GDP and accounts for a quarter of Swaziland's exports.

Since the 1990s the Swaziland economy has been increasingly dependent upon agriculture-based manufacturing, particularly in the capitalist estate sector. The intensification of foreign involvement accompanied changes in the structure of political power after 1968. The establishment in 1964 of the Matsapha Industrial Estate on the periphery of Manzini, the main commercial centre, set the stage for favourable development during the post-colonial era. Since then, industrial development in Swaziland has increased, helped by the priority it has received in the country's national development plans and an environment conducive to foreign investment created through a package of incentives designed to attract and encourage such investment. The benefits of development were numerous with increases in school enrolment, health facilities and other social services. Although rates of unemployment remained high, many jobs opened up with the growing numbers of factory establishments (Miles-Mafafo, 2001).

Industrial growth was interrupted by a severe recession during the late 1970s and early 1980s which drastically reduced employment and new growth. An upturn occurred linked to growing interest in Swaziland as a location for firms seeking to avoid South Africa in the face of anti-apartheid sanctions (Funnell, 1991). Sanctions against South Africa brought a wave of increased foreign investment, stimulated the stagnating economy and created a significant number of new jobs. However several transnational and local manufacturers in Swaziland also relocated their industries to South Africa's black 'homelands' to take advantage of the high incentives provided by the politically-inspired apartheid programme of industrial decentralisation. South Africa's protectionist trade policies, notwithstanding the Customs Union Agreement, also adversely affected Swazi efforts to attract new industrial concerns on several occasions, undermining infant Swazi industries including a television assembly plant and a fertilizer factory. Notwithstanding the Third National Development Plan's ambitious yet admirable objectives to promote rapid industrial growth, to indigenise and generate employment, at the end of the plan period in 1983 the objectives remained unfulfilled. South Africa's industrial decentralisation policy and Swaziland's inability to compete with the homelands for foreign investment together caused a severe decline in industrial growth, employment and real incomes.

Despite these setbacks, for nearly two decades, Swaziland has maintained high rates of economic growth and acquired one of the highest standards of living in sub-Saharan Africa due mainly to the relationship it holds with South Africa, and to the various regional institutional arrangements that back it (Oyowe, 1994). The processing of agricultural and forest products, especially sugar cane and wood pulp, has hitherto been the basis of industrial development in Swaziland. Wood pulp has until recently been Swaziland's second largest export, retaining a comparative edge over its Scandinavian rivals (Oyowe, 1994). Apart from these processing industries, other manufacturing activities operate on a comparatively small scale and largely consist of establishments producing light consumer goods and building materials.

Geographically, industrial development in post-colonial Swaziland has been confined to the country's main industrial area of Matsapha. Other industrial estates have been established in the south at Nhlangano, and in the north at Ngwenya. Proximity to major rail, road and air transportation routes to South Africa has favoured the rapid growth of Matsapha. Currently, the Matsapha Industrial Estate is highly diversified in terms of scale of units, products, services and types of ownership. The most significant national interest in manufacturing is held by Tibiyo takaNgwane, a Swazi national development fund and investment corporation. This agency operates a diverse portfolio of holdings in companies such as the Swaziland Meat Industries, Swaziland Brewers, Swaziland United Transport and Jubilee Printing and Publishing. Transnational interests represented at the Matsapha Industrial Estate include Coca Cola, Cadbury, Bromor Foods, Shell, YKK Zippers (a Taiwanese enterprise), Sikanye Shoes (Bata Shoes) and Natex (National Textile Industries). Macmillan Publishers have also made Matsapha the headquarters for their southern African operations (Stephen, 1986).

Also of note is the role of livestock and meat processing in the manufacturing sector. Livestock plays an important role in the lives of Swazis economically, socially and otherwise, but the recent outbreak of foot and mouth disease presented serious setbacks in terms of the loss of foreign exchange earnings as a result of the ban on national meat exports to the European Union (Madonsela and Dlamini, 2001).

Manufacturing in Swaziland not surprisingly shares some features in common with that of Lesotho (chapter 11). Employment in much of its industrial sector is largely female, for not dissimilar reasons. Swaziland is also a potential beneficiary from the new industrial export opportunities, particularly in garments and textiles production, heralded by the US Africa Growth and Opportunity Act.

Tourism

Historically, tourism has been vital to the Swazi economy. An important aspect was the expansion of casinos which began to open in the 1960s to attract South African tourists during the apartheid period (Crush and Wellings, 1987). Since the mid-1970s peak, the tourism economy has stagnated. With the restructuring of the South African casino industry in the post-apartheid period and the opening of new casino resorts in South Africa itself Swazi casino resorts have experienced a marked downturn (Harrison, 1995). Nevertheless, tourism continues to play an important role in the national economy with the majority of tourists drawn from South Africa.

The recent poor performance of tourism has implications for many small-scale enterprises such as the handicraft industry, which is dominated by women (Government of Swaziland, 2001). In terms of future development, new potential is offered by Swaziland's growing involvement in the Maputo Development Corridor Initiative (Rogerson, 2001). Planning for this cross-border development initiative has included a number of new tourism projects in Swaziland focussed around ecotourism and exploiting the biodiversity potential of the country. Opportunities for growth may develop through cross-border tourism initiatives which are being developed with South Africa's Mpumalanga province. An important element of the local economic development programmes being implemented in Manzini is also focussed on tourism promotion. Overall, the future prospects for tourism in Swaziland will be linked to the success of new efforts designed to promote tourism across southern Africa as a whole.

Structure of Investment

Since independence, changes in the structure of political power have been accompanied by an intensification of foreign capital involvement in all sectors, including agriculture, manufacturing, mining and tourism. In particular, South African capital is increasingly significant within the overall portfolio of foreign investment. This South African domination is increasingly evident even in agricultural and forestry enterprises that have long been funded by British private capital, Lonrho, and public funding through the Commonwealth Development Corporation (Rogerson, 1993).

As dependence on foreign capital intensified, the country's economy moved swiftly towards monocultural dependence on a single crop, sugar, for foreign exchange earnings (Daniel, 1986). Despite high levels of dependence on foreign capital, efforts were made to increase Swaziland's economic autonomy in the post-independence period. Although not entirely successful, through the establishment in 1968 of Tibiyo takaNgwane ('the wealth of the nation'), which has acted as a major vehicle for domestic capital accumulation, inroads have been made into foreign investment holdings. As a major shareholder in most of the country's investments, Tibiyo has secured the stakes of the Swazi aristocracy in a 'dependent, peripheral economy with a limited potential for domestic capital formation' (Daniel, 1986, p. 105). Through the King's custodianship over Tibiyo held in trust for the nation, a close alliance is maintained with foreign capital (Booth, 1983; Funnell, 1991; Atkinson, 1992).

British capital traditionally dominated the banking sector and large-scale capitalist agriculture, but has since been challenged by South African monopolies. Largely South African or South African-based investments fuelled growth in other sectors of the economy. In the manufacturing and commercial sectors, the most active investor has been the emerging South African conglomerate, Kirsch Industries, which dominates maize milling and importation, motor vehicle franchises, wholesaling, hardware and agricultural stores and also owns 50 per cent of the two major shopping malls. Other important South African conglomerates maintain manufacturing subsidiaries in Swaziland. Large-scale commerce and tourism are similarly dominated by South African capital (Atkinson, 1992). This domination of the Swazi economy by South African capital has in some respects been institutionalised by the operations of the Southern African Customs Union (chapter 15).

Swaziland's finance minister has emphasised the continuing importance of attracting foreign investment to underpin economic growth (Government of Swaziland, 2001). However, it is important when evaluating foreign direct investment, to analyse the nature of the jobs created and the conditions under which the workers find themselves. It is not uncommon for industries to exploit their workers by paying them low wages and subjecting them to hazardous working conditions with little or no protective clothing. The textile industries, for example, employ mainly women as casual labourers with no job security and very inflexible hours of work. It is important for Swaziland to maintain harmonious industrial relations not only for purposes of promoting foreign investment but also for the smooth operation of small- and medium-scale industries generated from domestic capital (Madonsela and Dlamini, 2001).

Whilst foreign investment does contribute positively to economic growth and employment, it is important to note that profits are largely repatriated to the countries of capital origin (Madonsela and Dlamini, 2001). It is therefore advisable for government, whilst strengthening the promotion of foreign investment, also to make a concerted effort to promote domestic investment. Whilst Madonsela and Dlamini (2001) applaud initiatives such as the E44m. fund directed at promoting small- and medium-scale enterprises, they also caution the government to ensure that projects financed using these funds should be of a sustainable nature and have the potential to regenerate the fund to provide new opportunities for the Swazi people.

Another problem that can arise due to heavy reliance on foreign investment is that government revenues may end up stagnating due to the pressure to offer generous tax concessions. In the past Swaziland has experienced a problem of 'fly by night' companies which decide to relocate to other countries as soon as the tax holiday is over or change their status and start operating under a new name.

Recent Social and Economic Reforms in Swaziland

Swaziland faces a number of social and economic challenges in the context of globalisation and a situation of peripheral dependence (Miles-Mafafo, 2001). Responses have included a recent set of reforms which include the finalisation and launching of the National Development Strategy, the development of the Poverty Reduction Strategy and Action Plan, the development of a population policy and public sector reform measures. These initiatives address themselves specifically to the serious problems of poverty, population growth (2 per cent p.a.), poor delivery of the public sector and the issues surrounding HIV/AIDS (Madonsela and Dlamini, 2001). This concluding section reviews several of these initiatives.

Economic Growth, Distribution and Poverty Reduction

One of the goals of Swaziland's national development strategy is the acceleration of economic growth to ensure a sustainable upliftment of the quality of life for the Swazi people. However, whilst increased economic growth would definitely be a significant step towards this objective, the income distribution programme still remains very important in the attainment of sustainable economic development. The extent to which increased production converts into an instrument of increased welfare for society depends on equitability/adequacy of distribution measures. For example, as long as women and children remain in the periphery of the distribution process, positive economic growth cannot be expected to improve standards of living (Madonsela and Dlamini, 2001). In the 2001 budget speech it was stated that financial and technical assistance for the development of a poverty reduction strategy has been solicited from the World Bank (Government of Swaziland, 2001). Such a programme needs to be accompanied by a sound population policy in which incentives designed to reduce family size are a priority.

Millennium Projects

The intentions of the government of Swaziland for the next two decades include the introduction of millennium projects aimed to improve the tourism industry and to promote investment and job creation (Government of Swaziland, 2001). One such project, the construction of a new hotel/sporting complex, has obvious attractions but raises questions when many existing hotels are on the verge of closing down due to poor maintenance, seasonality of demand and other problems (Madonsela and Dlamini, 2001). Another millennium project is the construction of a new international airport. Critics point out that Swaziland's existing airport is already under-used and massive retrenchments in the airline industry have occurred since its takeover by South African Air Link.

Infrastructural Development

Significant reform has been made in Swaziland in terms of infrastructural development. The Micro Projects Scheme aims at development of community infrastructure, which is a positive move towards the attainment of better living standards in rural communities. For this project, E40m. has been allocated for regional development, which it is anticipated will go a long way into reducing regional spatial inequities as well as curbing rural-urban migration flows (Madonsela and Dlamini, 2001). Through the improvement of infrastructure in rural communities, investment flows will be redirected from urban areas in an attempt to create rural employment opportunities, thus reducing pressure on an already congested urban environment. Women are seen as direct beneficiaries of increased rural incomes deriving from such rural development programmes. Rural development is important for fostering healthy family relations. With reduced rural-urban migration flows, family disintegration which in most cases was a direct consequence of spouses being forced to leave their homes in search of employment could be reduced (ibid.). Rural development is also needed to reduce inequities in education and health.

Significant efforts have been made in the development of infrastructure particularly in the improvement of the road networks. The construction of the Manzini-Mbabane highway, improvements made on secondary roads and the continuing process of upgrading and improving roads is evidence of the commitment to improve transport and communication.

One infrastructural development that has come under intense criticism is the Maguga Dam Project costing E450m. It concentrates mainly on promoting irrigation activities by smallholder sugar cane farmers, but its sustainability has been questioned given the volatile prices of sugar in the world market. The government has also committed itself to undertake similar work in the Lower Usuthu basin. This dam project is estimated to cost approximately E701m. Concern has been expressed about the undertaking of such an expensive project when the Maguga Dam project itself is still incomplete and its viability has not been proven. It is likely to perpetuate the budget deficit and increase government debt.

Privatisation

In response to external pressures to compete in the global market and to increase the efficiency of public enterprises, the government is embarking on a privatisation programme. This has the potential to improve both the efficiency and sustainability of current public institutions such as the Swaziland Development and Savings Bank, although it may have negative implications for consumers if the prices of goods and services increase.

Conclusion

This chapter has focussed upon the major features of Swaziland's changing economic geography in order to understand the contemporary challenges that confront this African micro-state. For Swaziland, there are many challenges posed by globalisation and the changing political economy of the wider southern Africa region. The strength of linkages to South Africa continues to dominate the economic landscape of the country. The growth of cross-border development initiatives offers promising opportunities for potential economic revival. Local efforts are also being made towards redressing many of the economic and social ills that continue to plague the country's uneven trajectory of social and economic development.

References

Atkinson, C. (1992), *Regional Industrial Change in Southern Africa: a case study of Swaziland in the 1980s*, unpublished MA Dissertation, Department of Geography, Queens University, Kingston, Canada.

Bonner, P. (1983), *Kings, Commoners and Concessionaires: The Evolution and Dissolution of the 19th Century Swazi State*, Ravan, Johannesburg.

Booth, A. (1983), *Swaziland: Tradition and Change in a Southern African Kingdom* Westview, Boulder.

Crush, J. and Wellings, P. (1987), 'Forbidden Fruit and the Export of Vice: Tourism in Lesotho and Swaziland', in S. Britton and W.C. Clarke (eds), *Ambiguous Alternative:*

Tourism in Small Developing Countries, University of South Pacific, Suva, pp. 91-112.

Daniel, J. (1986), 'Swaziland', in J. Hanlon (ed), *Beggar Your Neighbour: Apartheid Power in Southern Africa*, James Currey, London, pp. 91-106.

Esterhuysen, P. (ed) (1998), *Africa A-Z Continental and Country Profiles*, Africa Institute, Pretoria.

Fair, T.J.D. (1981), *Towards Balanced Spatial Development in Southern Africa*, Africa Institute, Pretoria.

Funnell, D. (1991), *Under the Shadow of Apartheid: Agrarian Transformation in Swaziland*, Avebury, Aldershot.

Government of Swaziland (2001) *Budget Speech 2001*, Ministry of Finance, Mbabane.

Harrison, D. (1995) 'Development of Tourism in Swaziland', *Annals of Tourism Research*, vol. 22, pp. 135-56.

Madonsela, W. and Dlamini, B. (2001), 'A Critical Analysis of the Swaziland Budget Speech 2001 for the Welfare of Women and Children', unpublished paper prepared for the Council of Swaziland Churches Department of Justice Peace and Reconciliation.

Miles-Mafafo, M. (2001), 'Fantasies of Development and Housing Provision for All?: Revisiting Urbanization Issues in Swaziland', *Urban Forum*, vol. 12, pp. 105-18.

Oyowe, A. (1994), 'Country Report: Swaziland', *The Courier*, vol. 147, pp. 18-38.

Rogerson, C.M. (1993), 'British Aid to Southern Africa: the Role of the Commonwealth Development Corporation', *Africa Insight*, vol. 23, pp. 209-18.

Rogerson, C.M. (2001), 'Spatial Development Initiatives in Southern Africa: the Maputo Development Corridor', *Tijdschrift voor Economische en Sociale Geografie*, vol. 92, pp. 324-46.

13 Zimbabwe: In Search of Spatial and Social Equity

LOVEMORE ZINYAMA AND DANIEL TEVERA

Introduction

Zimbabwe has been going through an economic and political crisis since the late 1990s. The purpose of this chapter is to explain the politico-social and economic factors underlying the crisis. The country attained political independence in 1980 after 90 years of colonial rule under an Anglo-European minority government. Independence was achieved after a decade and a half of civil war between the African liberation movements on one hand and the settler European government on the other. Both colonial and post-colonial policies have left an indelible imprint on the present-day geography of the country.

The Resource Base

Population

Since the early 1960s, Zimbabwe has experienced a high rate of population growth, above 3.0 per cent per year, with a doubling time of twenty years. However, for reasons that will be explained later, recent demographic surveys indicate a decline in the population growth rate.

The population of about 12m. is unevenly distributed across the country. Some two-thirds of the population resides in rural areas, which fall into three main categories based on the type of land tenure and farming systems. The communal lands (Figure 13.1), designated for Africans during the colonial era, are characterised by subsistence smallholder agriculture. At the last census in 1992, the communal lands held 51 per cent of the national population, with people living in relatively dense village settlements. Land that was assigned for European settlement during the colonial period is dominated by large-scale commercial farms, some of them as much as 5,000 hectares in size, with generally dispersed settlement. In 1992, the commercial farming areas held 13 per cent of the population. The third type of rural settlement occurs in the post-independence resettlement schemes that have been implemented by the government under the land reform and redistribution programme. The resettlement areas had about 4 per cent of the national population in 1992.

Figure 13.1 Zimbabwe

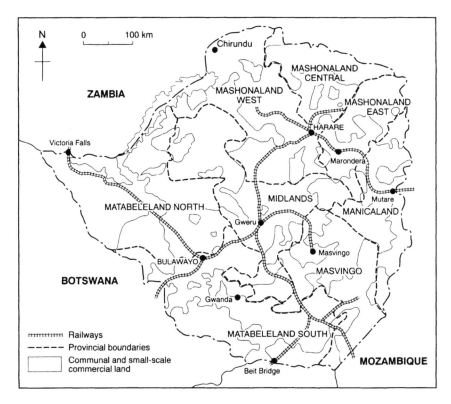

The average national population density is approximately thirty persons per km^2. However, communal lands are the most densely populated, with some areas having densities as high as sixty persons per km^2. Large-scale commercial farming areas are the least populated, with an average density of ten persons per km^2 (Tevera, 1994).

Internal migration, both before and after independence, has had a major impact on the distribution of the population. Since independence, urban areas, particularly the large cities, have experienced a large influx of rural migrants seeking employment. In the rural areas, the state-sponsored land redistribution programme has also led to major shifts in population, as some 70,000 peasant families were transferred from the overcrowded communal lands to former sparsely populated commercial farmlands.

Zimbabwe has also experienced different forms of international migration including European settler migrations during the colonial period, refugee movements by Zimbabweans into neighbouring countries during the 1970s war and by Mozambicans during that country's civil war in the 1980s, labour migration in search of employment in neighbouring countries, as well as undocumented migration in and out of the country both before and after independence (McDonald, 2000; Zinyama, 1990; Zinyama and Zanamwe, 1997). However, international migrants account for only a tiny proportion of the national population – less than five per cent of the population at the 1992 census had been born outside the country. For a long time, Europeans either entering or leaving the country accounted for the bulk of the international migration streams, the volume and direction of movement depending on changing domestic political conditions (Zinyama, 1990). But more recently, a growing number of black Zimbabweans - professionals and other skilled workers as well as unskilled people - have been leaving to seek employment in countries such as South Africa, Botswana, the United Kingdom and Australia in order to escape the deteriorating economic conditions at home. While professionals and other skilled people generally leave the country legally through official crossing points, large numbers of unskilled and poorly educated people cross the border illegally into neighbouring Botswana and South Africa in search of employment. The numbers of such illegal migrants are unknown, but the Zimbabwean High Commission in Pretoria estimated that as many as 460,000 Zimbabweans, including some 60,000 professionals, were working in South Africa in the late 1990s (Fultz and Pieris, 1998).

HIV/AIDS is the biggest public health and development challenge facing Zimbabwe today. The epidemic has wreaked social and economic havoc in the population during the past decade. It is currently estimated that 20-25 per cent of the adult population is HIV-positive. Over 1,200 people are said to die of AIDS every week, a majority of them in the economically active age groups 20-40 years. The number of children under 15 years living without one or both parents now exceeds 1 million, or approximately one-twelfth of the total population. The epidemic has imposed severe social and economic pressures on the population, lowered labour productivity, and placed a major strain on the country's poorly funded and under-staffed health delivery system.

Water Resources

Zimbabwe is relatively well endowed in terms of surface and groundwater resources, although supplies are unevenly distributed geographically. The central and eastern parts of the country receive higher rainfall than the southern and western regions. The country receives an average of 675 mm. of precipitation per year, but rainfall is highly variable from year to year and droughts are a recurrent phenomenon (Mazvimavi, 1998). High evapotranspiration accounts for the low proportion that reaches the country's rivers and other water bodies as runoff. Potential evapotranspiration is highest in the hotter lower-lying regions in the south and west of the country.

There are no natural lakes, but wetlands or *dambos* on the central highveld are important water sources for drinking and small garden irrigation by smallholder farming families in the communal lands, especially during the dry season. In addition to these natural surface water bodies, Zimbabwe has about 10,600 dams, the vast majority only small reservoirs (including some 5,700 dams on commercial farms) with a capacity of less than 1m. m^3 each (Zinyama, 1995). Surface water resources contribute 90 per cent of the country's total water supply (Chenje et al., 1998).

Groundwater supplies are estimated at 2,000m. m^3 per annum (Grizic, 1980). Groundwater could be a major source of water for the country, but it is under-utilised at the moment, mostly for irrigation on commercial farms. The Nyamandhlovu aquifer in the west of the country has proved to be an important additional source of water for the drought-prone Matabeleland region and the City of Bulawayo. The perennial problem of water shortage in Matabeleland has given rise to the proposed multi-million-dollar Matabeleland Zambezi Water Project (MZWP). The project, which has been under consideration since the early twentieth century, envisages the construction of a 450km. pipeline which will draw water from the Zambezi River upstream of the Victoria Falls to Bulawayo.

On aggregate, the country appears to have adequate water resources. However, water supply often fails to meet demand particularly during the dry season and in drought years. This is the case partly because the resource is not fully exploited. The shortages may result in struggles over access to the resource (Swain and Stalgren, 2000). Water conflicts at local, provincial and national levels manifest themselves in several ways, sometimes culminating in physical confrontation. Irrigation agriculture is by far the biggest user of water, consuming about 70 per cent of the total water used in the country, mostly in the commercial farming sector. By the mid-1990s, the latter sector accounted for 94 per cent of the total irrigated area of 14,890 hectares in the country (Chenje et al., 1998; Mazvimavi, 1998). The total demand for water is increasing in line with the increasing urban population, and the expansion of mining, manufacturing and agricultural activities.

Wildlife and Forests

Zimbabwe has diverse tropical fauna and flora. However, the country's natural habitats are being degraded and biodiversity is under threat. Deforestation is a major problem, especially in the communal lands and resettlement areas. It is estimated that during the late 1980s as much as 70-100,000 hectares of woodland was being cleared annually for various purposes such as agriculture, settlement, construction materials and for medicinal purposes (Whitlow, 1988). Less than 10 per cent of the total land area remains under closed canopy or plantations (SADC-ELMS, 1997). Only 2.4 per cent of the land area is afforded legal protection as state forest under the management of the Forestry Commission.

Today, much of Zimbabwe's wildlife is concentrated in the country's national parks, safari areas, recreational parks, sanctuaries and botanical gardens, of which Hwange National Park (1.5m. hectares) is by far the largest. Most parks and wildlife areas are located in areas of low rainfall, sparse population and poor agricultural potential along the country's borders (Heath, 1992). The Department of National Parks and Wildlife Management (DNPWM) is responsible for the management of the 5m. hectares of the Parks and Wildlife Estate which account for 13 per cent of Zimbabwe's total land area. The department is also responsible for the conservation and sustainable productive use of indigenous plant and animal resources in communal, commercial farming and resettlement areas of the country. In the early 1980s, the department and other stakeholders undertook to involve local communities in the management of indigenous resources as a deliberate strategy in the fight against the destruction of wild animals and flora. Communities in a growing number of communal lands are participating in, and benefiting from, the local wildlife through the Communal Areas Management Programme for Indigenous Resources (CAMPFIRE) (Heath, 1992). The programme seeks to place responsibility for the management of indigenous natural resources under the control of resident communities, who in turn will derive direct benefits from those resources.

The Parks and Wildlife Estate is the major tourist attraction which earns the country much needed foreign currency. Activities available to tourists include game viewing, hunting and non-consumptive safaris, white-water rafting, canoeing, angling, hiking and mountain climbing. In addition, since the 1980s, game ranching has become a major economic enterprise on commercial farms, particularly in the drier western and southern regions of the country. The farmers run photographic and hunting safaris for tourists. Profits per hectare from game farming are considerably higher than those from cattle ranching in these agriculturally marginal areas (Taylor, 1988).

Minerals

Zimbabwe has a rich and diverse mineral base and produces over forty minerals on a commercial basis. Of these, gold, asbestos, nickel, coal, copper, chrome and tin account for 90 per cent of the total value of output. In terms of occurrence, most of the minerals are found on the central plateau along a highly mineralised belt stretching from north-east to south-west known as the Great Dyke.

The mining sector accounts for less than 10 per cent of formal employment but contributes 30-45 per cent of the country's total annual export earnings. Foreign-owned companies such as Rio Tinto, the Anglo American Corporation and Lonrho dominate the sector. However, the government has also been a major player in the mining sector since the early 1980s, both as an operator through state-owned mining parastatals such as the Zimbabwe Mining Development Corporation, and through its control of minerals marketing. There are also numerous small mines owned by individuals, small companies and co-operatives. A substantial but unknown amount of illegal alluvial panning for gold also takes place, much of it by the rural poor as one of their strategies for coping with poverty and food shortages, especially during droughts. The number of panners fluctuates by season, availability of food and prevailing economic conditions. For instance, it was estimated that the number of alluvial gold panners during the 1991-92 drought swelled to as many as 200,000 (Ngara and Rukobo, 1999).

The Colonial Economy

Land alienation for European settlement that started after colonisation in 1890 has left an indelible imprint on the political, economic and social geography of Zimbabwe. Throughout the colonial period from 1890 to 1980, the European population never exceeded more than 5 per cent of the total population of the country. Even at its peak in the 1960s, only 228,296 Europeans were enumerated at the 1969 census. Yet Europeans had exclusive rights to a large proportion of the country allocated to them by the colonial administration, reaching a peak of 53 per cent of the total country in the early 1960s, and still 41 per cent at independence, encompassing the most productive areas. The indigenous people who were forcibly removed from their traditional lands along the central highveld of the country were resettled in sparsely populated, drier, agriculturally less suitable, and sometimes tsetse fly infested, regions away from the central watershed. Others were moved into progressively overcrowded and overstocked native reserves, where severe environmental degradation became rampant (Chenje et al., 1998; Tevera, 1994; Whitlow and Zinyama, 1988; Zinyama and Whitlow, 1986).

Not surprisingly, during the war for independence in the 1970s, land hunger was the major rallying point for the African liberation movements. After independence, the new African majority government was expected to effect major land redistribution from the European landowners. Land redistribution was essential for a number of reasons, notably:

- to relieve population pressure in the crowded former African reserves, now known as communal areas;
- to reduce environmental degradation;
- to give land-hungry African peasants an opportunity to venture into small-scale commercial agricultural production for the first time.

Two decades after independence, land redistribution remains an unresolved and increasingly violent aspect of the economic, social and political geography of Zimbabwe (see below).

The apportionment of land, by determining where each racial group could reside and on what terms, greatly influenced the patterns and processes of urban growth and urbanisation. For the first few decades of the twentieth century, the economy was dependent on agriculture. Significant industrial growth commenced during World War II (1939-45) when the country faced shortages of imported manufactured goods from a war-ravaged United Kingdom, its principal trading partner. Industries established during this period include the country's sole iron and steelworks at Redcliff near Kwekwe, cotton spinning and weaving, and clothing factories. Industrialisation received another boost from the late 1960s to the mid-1970s when the government actively promoted import substitution industrialisation in order to obviate shortages of imported goods following the imposition of international economic sanctions by the United Nations against the illegal white settler-colonial regime in the mid-1960s.

Although industrialisation was taking place from the 1950s, the politics of racial separation restrained the urbanisation of the African population until the late 1970s. Influx control laws barred Africans from owning or leasing land in urban areas or moving into town as permanent residents. Africans could only stay on their employers' premises (if they were employed as domestic workers) or rent accommodation in the townships from the urban local authorities. But they were allocated rented municipal housing only during periods of employment. Local authorities were required to register all employed Africans and persons coming to visit from the rural areas and to determine the visitors' length of stay in town. The insecurity of urban living for Africans was compounded by low wages inadequate to support a family, together with the lack of social and financial security in times of old age or unemployment. Consequently, it was mostly males who came to work in town for limited periods of time, leaving their wives and children to farm in the rural areas. Rural-urban migration by the African population only accelerated in the late 1970s as the security situation in the rural areas deteriorated during the war for independence. Large numbers of rural people fled to the towns while urban workers were forced to bring their families into town for their safety.

Until the early 1970s, less than 20 per cent of the population was urbanised. At the 1982 census, two years after independence, the proportion of the population enumerated as urban had increased to 27 per cent, rising to 31 per cent by 1992. It is estimated that some 45 per cent of the national population was urban by 1999. The number of places classified as urban more than doubled from 34 in 1969 to 76 in 1992.

The Transformation of Economy and Society, 1980-1990

At independence in 1980, the new African majority government set out to transform the economy and society in order to redress the socio-political and racial disparities of the colonial era. The government embarked on a number of policy initiatives that, at the time, were underpinned by its drive to establish a socialist state in the country. A review of government development policies over the past twenty years shows a clear disjuncture around 1990-91. The first ten years from 1980 were strongly influenced by the ruling party's adherence to socialism. But, by the end of the decade, it had become apparent that, while significant gains had been achieved especially in the social sector, major economic constraints remained unresolved. These problems were largely responsible for the sluggish or 'stop-start-stop-start' performance of the economy, increasing unemployment and growing foreign exchange shortages. Because of the deteriorating economic conditions, the government was compelled to abandon its socialist policies and turn to the International Monetary Fund

(IMF) and the World Bank for balance of payment support while it endeavoured to restructure and liberalise the economy. The result was the adoption, in 1990-91, of what became known as the Economic Structural Adjustment Programme (ESAP). This became the government's principal policy platform for the next five years. But ESAP generally failed to address the macro-economic problems facing the country. The economy continued to deteriorate throughout the 1990s, with worsening unemployment, high inflation, flight of capital investment and de-industrialisation, failure to attract new foreign investment, and deteriorating social services. The economic problems have been compounded by the HIV/AIDS pandemic that is placing enormous pressures on social services, especially health services, while taking a heavy toll on the most economically productive segment of the population.

The new government quickly embarked on programmes to improve access to economic and social services for the previously poorly serviced African population in both urban and, especially, rural areas. In the area of social services, the focus was on the provision of educational and health services, and the rehabilitation of infrastructure that had been destroyed during the war in the 1970s. Free medical services were provided for low-income and unemployed groups in both rural and urban areas. A hierarchy of health facilities was established, ranging from rural clinics through district and provincial hospitals, up to central hospitals in Harare and Bulawayo that were to serve as national medical referral centres. Other interventions aimed at improving the health status of the population included water supply and sanitation programmes for rural communities; maternal and child health care, including child immunisation against the five child killer diseases (measles, whooping cough, diphtheria, poliomyelitis and tetanus); and an extensive community-based family planning programme and contraceptive distribution network (World Bank, 1989).

Likewise, the state introduced free primary education for all children up to approximately 13-14 years of age. The number of primary schools increased from 3,161 at independence in 1980 to 4,539 by 1990, with a majority of the new schools in the previously poorly serviced rural areas (Republic of Zimbabwe, 1983, 1991). More significantly, the number of secondary schools increased almost sevenfold from 197 in 1980 to 1,512 in 1990. Primary school enrolment increased by 72 per cent from 1.24m. pupils in 1980 to 2.12m. by 1990 while secondary school enrolment increased eightfold from 74,320 pupils to 672,650 during the same period.

Probably the most remarkable progress was achieved in the transformation of the smallholder agricultural sector in order to raise rural incomes and living standards. The government embarked on several programmes to support communal area farmers by providing them with inputs, extension services, credit facilities, marketing and transport for their produce such as maize, cotton and groundnuts (Rukuni and Eicher, 1994; Zinyama, 1986). The response from the farmers was so tremendous that Zimbabwe became self-sufficient in its basic food requirements under normal weather conditions and soon acquired the status of a regional breadbasket for southern Africa.

By the end of the first decade after independence, Zimbabwe's success in the provision of social services was widely acclaimed by international organisations and the donor community. Literacy rates had improved significantly, the proportion of children (including girls) of school-going age who were actually in school had increased, infant mortality had dropped while life expectancy had gone up. The total fertility rate, which in 1988 stood at 5.5 children per woman in the 15-49 age groups, declined to 3.9 by 1999 (CSO and Macro International, 1995). The population growth rate was showing signs of declining by the end of the decade, thanks in part to improved literacy especially among women, and in part to the widespread use of modern family planning methods among females of child-bearing age (World Bank, 1989).

However, these impressive gains in the provision of social services were placing an enormous financial burden on the exchequer. By the end of the 1980s, it was apparent that the social policies of the first decade were no longer financially sustainable. Hence, from the early 1990s, there was a change of policy with the introduction of cost recovery measures and user fees for services rendered as part of ESAP.

At independence, Zimbabwe had one of the most developed and diversified economies in sub-Saharan Africa, second only to South Africa, supported by a relatively sophisticated infrastructure. Manufacturing accounted for approximately one-quarter of the gross domestic product. However, this modern economy was grossly iniquitous, both spatially and racially. Europeans and multinational corporations dominated the key sectors of the economy, such as commercial agriculture, manufacturing, mining and commerce. Foreign penetration of the economy had been particularly marked during the UDI years (1965-1979) when the colonial government actively promoted inward investment to counter international economic sanctions. An estimated three-quarters of manufacturing capacity was in the hands of foreign companies, principally South African, but also British and American corporations acting through their South African subsidiaries.

In keeping with its early socialist ideology to redress the racial inequalities in the modern economy and to reduce the extent of foreign ownership (particularly South African), the new government adopted several measures that sought to bring about greater domestic control of the economy. It did so through state ownership during the 1980s and through African economic empowerment in the 1990s (Herbst, 1990). The principal vehicle used by the government to achieve local ownership of foreign enterprises in key sectors of the economy was through the establishment of parastatal institutions that were to serve as the state's holding agencies (Zinyama, 1989). The concept of parastatals was not entirely new in Zimbabwe. During the UDI period, the colonial regime had also established several parastatal organisations to rescue loss-making firms that were threatened with closure because of the difficulties of trading under international economic sanctions. The parastatals were also tasked to promote import substitution industrialisation by producing goods that had previously been imported from abroad. One such parastatal that played a key role in sustaining the colonial economy during this period of international isolation was the Industrial Development Corporation (IDC), formed in 1963. Even today, the IDC continues to serve as the major vehicle through which the state participates in the manufacturing sector. It is a major shareholder in several sub-sectors including motor vehicle assembling, cement manufacturing and mineral processing, chemicals and explosives, fertilizers, glass, clothing and packaging. Another parastatal that was in existence long before independence is the Zimbabwe Iron and Steel Company (ZISCO), operator of the country's only steelworks at Redcliff since the mid-1940s.

During the 1980s, the government established several new parastatals or extended the responsibilities of existing ones. These include the Small Enterprises Development Corporation which was tasked with promoting and financing small- and medium-scale enterprises; the National Oil Company of Zimbabwe which has been responsible for the procurement of refined petroleum products; the Minerals Marketing Corporation of Zimbabwe which was responsible for the marketing of all minerals and metals with the exception of gold; and the State Trading Corporation which was responsible for the importation of household consumer goods and various other items prior to the liberalisation of the economy in the early 1990s. The role of the Zimbabwe Tourism Development Corporation was expanded from tourism promotion to include the management and ownership of hotels. The Zimbabwe Mass Media Trust was established shortly after independence to take over control of the main national newspapers and other publishing businesses from South African interests.

By the late 1980s, many parastatals became heavily dependent on state subsidies. There were also allegations of poor management, excessive government interference in their day-to-day operations, corruption and nepotism in their employment practices and in the awarding of contracts.

These problems, coupled with the new IMF/World Bank ethos that government should not be involved in commercial enterprises, have led to moves to privatise or commercialise several of the parastatals. A few have already been privatised, for instance the Cotton Marketing Board which became the Cotton Company of Zimbabwe, Dairibord Zimbabwe (from the former Dairy Marketing Board), and the Commercial Bank of Zimbabwe (formerly BCCI). The Zimbabwe Tourism Development Corporation was disbanded and its hotel business activities were taken over by a new company, Rainbow Tourism Group. All four examples are now public companies quoted on the Zimbabwe Stock Exchange. Another reason for selling the parastatals is that the government desperately needs the money to finance the large national budget deficit.

There have also been significant shifts in the area of international trade relations between the 1980s and the 1990s, mainly arising from the country's changing relations with South Africa. During the 1980s, before the advent of a new democratic dispensation in South Africa, the independent countries of southern Africa sought to disengage themselves from their militarily and economically more powerful southern neighbour. In the case of Zimbabwe, however, disengagement during the 1980s was largely unsuccessful because of the country's high dependence on both South African markets and transport routes for access to the sea and overseas trading partners. South Africa remained Zimbabwe's principal trading partner, followed by the United Kingdom and Germany.

Economy and Society, 1990 to the Present

The decline in economic growth and living standards that had started in the late 1980s persisted throughout the 1990s. The 1990s were dominated by government's largely unsuccessful efforts to implement the economic structural adjustment programme. ESAP has widely been blamed for causing and intensifying poverty amongst the country's population. By the end of the 1990s, the programme had, to all intents and purposes, been abandoned as too costly in social, political and economic terms. Zimbabweans are poorer now than at independence in 1980, with an estimated three-quarters of the population living below the poverty datum line. In the latter part of the 1990s, other, politically less costly, macro-economic strategies were attempted by the government, notably (a) economic indigenisation in an effort to redress past imbalances in the ownership of the means of production, and (b) a renewed vigour in the implementation of land reform.

The Economic Structural Adjustment Programme

The structural adjustment programme adopted in 1991 fell into four broad categories:

- fiscal policy reform which aimed to reduce the government budget deficit, reduce the civil service, and eliminate subsidies to loss-making parastatals;
- monetary policy reform which, among other things, deregulated the foreign exchange markets and liberalised the financial sector;
- trade policy reform which resulted in the opening up of the trade regime for importers and exporters;
- regulatory policy and institutional reform which sought to simplify the approval process for private investment and to simplify labour regulations.

Unfortunately, the government persistently missed the targets that were agreed with the World Bank/IMF, resulting in the suspension of balance of payment support to the country in 1997.

Following the suspension of ESAP, the government chose to pursue what it considered a more home-grown reform programme that would be sensitive to local economic and social conditions. The Zimbabwe Programme for Economic and Social Transformation (ZIMPREST) was launched in 1997-98. Like its predecessor, ZIMPREST aimed to promote economic growth and social development through:

- restoration of macro-economic stability;
- promotion of private sector savings and investment;
- indigenous economic empowerment;
- poverty reduction, employment creation and entrepreneurial development;
- investment in human resources development;
- provision of safety nets for vulnerable groups.

Removal of educational and health subsidies that had been available for the poor during the 1980s, deregulation of food prices and removal of food subsidies, and retrenchments in both the public and private sectors have hurt the poor and other vulnerable groups the most (Mlambo, 1997; Tevera, 1995, 1997). In the education sector, government real expenditure per child declined by about 17 per cent between 1990 and 1992. Access to education for children of poor families has become increasingly difficult following the re-introduction of school fees. In contrast to the gains of the

1980s, the proportion of children of school-going age not actually attending school has increased during the past decade.

The provision of health services has also been adversely affected by the introduction of cost-recovery measures and user fees, while the annual budget for the Ministry of Health has shrunk in real terms during the past decade. More recently, severe shortages of foreign currency with which to import essential drugs and equipment have severely compromised service delivery. The public health sector has lost many professionals to the domestic private health sector or through emigration to South Africa, the United Kingdom and elsewhere as staff flee deteriorating working conditions. The inability of families to access health services may be illustrated by recent trends in mortality rates, and this has been compounded by the AIDS epidemic. The maternal mortality rate, which had dropped to 251 per 100,000 live births by 1991, has shown an upward trend during the past decade, rising to 283 in 1994 and 695 by 1999 (CSO and Macro International, 2000). The 1999 survey by the CSO and Macro International (2000) attributed the recent deterioration in child survival rates to increasing poverty and the direct and indirect impacts of the AIDS epidemic.

In the rural areas, communal farmers' access to agricultural inputs such as hybrid seeds and chemical fertilizers has deteriorated due to inflation-driven price escalations. At the same time, the government has been under pressure to reduce public expenditure. An inadequate budget for the Ministry of Agriculture means that extension workers are no longer able to carry out crucial activities such as farm visits and farmer training. The Agricultural Finance Corporation, which used to provide seasonal credit to smallholder farmers to purchase inputs, has virtually ceased doing so because of the high default rates in loan repayment. During the 2000 marketing season, the Grain Marketing Board was experiencing serious cash-flow problems such that it was unable to pay the farmers who had delivered their produce until after the start of the 2000-01 season. As a result, smallholder farmers were unable to purchase their inputs in time for the new season. Overall, there is growing vulnerability and poverty in the rural areas because of the withdrawal of many of the support services that farmers used to receive from the state.

In order to cushion themselves against the continuous deterioration in living standards, the poor, and even many urban middle-income households, have adopted a variety of coping strategies. These include informal sector activities, walking to work instead of using public transport, use of traditional medicines instead of more expensive modern medicines, and extensive urban cultivation of basic foodstuffs on either privately owned gardens or vacant municipal land (Drakakis-Smith et al., 1995;

Mudimu, 1996; Rakodi, 1994; Tevera, 1995). Low- and middle-income urban families with surplus living space have resorted to sub-letting rooms in order to supplement household incomes. In some cases, urban house-owners have even sold their houses in middle-class residential areas to take up accommodation in the overcrowded low-income suburbs or townships. Further testimony of the harsh economic conditions of the late 1990s is given by the massive expansion of the informal economy, particularly street vending, petty commodity production, waste scavenging and cross-border trading between Zimbabwe and its neighbours. Prostitution caused by poverty has also contributed to the HIV/AIDS epidemic. Ties with the rural areas through land cultivation and food transfers to town and through return migration when one becomes unemployed are being reinforced because of urban social insecurity (Potts and Mutambirwa, 1990; Tevera and Dahl, 2000).

Land Redistribution

The land redistribution programme began in 1981, a year after independence. At the onset, the government had planned to transfer enough land from European farmers to resettle some 162,000 peasant families by the mid-1980s. But by the end of the decade, only 53,000 families had been resettled on 3.3m. hectares of land. The government's current goal is to transfer a total of 8.5m. hectares from the large-scale commercial farming sector for redistribution to peasant farming families. In the end, the large-scale farming sector (now including a small number of African landowners) will be reduced from 15.5m. hectares at independence to only 5m. hectares, to be used mainly for the production of strategic and capital-intensive crops such as tobacco, wheat and horticultural products.

The principal reasons for the slow pace of land resettlement during the 1980s included both a shortage of financial resources for land purchase and development and the small amount of suitable land offered to the government for resettlement under the stringent 'willing-seller-willing-buyer' conditions enshrined for the first ten years in the constitution agreed at the Lancaster House Conference in 1979. There are also indications that, in the late 1980s, the government had lost some of the political will to implement the land redistribution programme. While the need for land redistribution was periodically brought up, in practice, the amount of money set aside in the national budget each year was never adequate for land purchase and the provision of support services and infrastructure. It was not until the 1990s that land redistribution assumed a new urgency and, more lately, has turned increasingly violent, with large numbers of ruling party supporters forcefully occupying European-owned farms in all parts of the country. Former nationalist freedom fighters and other landless peasants had become critical of the government's lack of urgency in implementing the reform programme. Illegal occupations of European-owned farms started in late 1997, spearheaded by the Zimbabwe National Liberation War Veterans Association (ZNLWVA), and intensified during 2000 and 2001.

The changes in legislation enacted during the past decade are crucial to understanding the current phase in the land reform programme. First was the amendment to the country's constitution in December 1990, removing the entrenched clauses that had protected property rights against compulsory acquisition during the first ten years after independence. This was followed by the Land Acquisition Act of March 1992 which allowed for the acquisition of land after appropriate judicial adjudication to determine levels of compensation. The passing of the 1992 Act was expected to speed up the process of land acquisition, but this turned out not to be the case. In 1998, following a UNDP sponsored donors' conference at which the international community made pledges to support the land reform programme under certain conditions, the government published a new national land policy framework. This framework launched the second phase of the land reform programme in June 1999. Further amendments aimed at making it possible for government to speed up the land reform programme were made to the Land Acquisition Act in 1999 and to the constitution in 2000.

The new legislation gave the government wider powers to acquire land for resettlement than hitherto. The current phase of the land reform programme, launched in June 1999 under the new provisions, envisages the resettlement of some 118,000 peasant families in the shortest possible time. The 2000 constitutional amendment made the former colonial power, the United Kingdom, ultimately responsible for compensation to commercial farmers whose land was expropriated for resettlement. This became a source of further public and often acrimonious disagreement between the two governments. The Zimbabwe government declared that it would pay compensation for improvements only (and payable over a period of up to five years depending on availability of funds), and not for land which, it argued, had originally been forcibly taken away from its people. For its part, the British government argued that it is willing to provide financial support amounting to some £36m. for the land reform programme, provided it is carried out in a transparent manner and with a focus on poverty alleviation. Of major concern to the British government, other governments in the region and the international community at large, has been the violence and apparent extra-judicial process that has accompanied land reform since the late 1990s.

Land reform has, since February 2000 when the government lost the referendum for a new constitution, become highly politicised and violent, with the government accusing commercial farmers of conniving with the main opposition political party to frustrate the programme. According to the government, the farm invasions are only a manifestation of the need for urgent land reform and resettlement of the country's landless majority. In response, international donors suspended virtually all development aid to the government in protest against the occupation of the farms and violence directed at the white farmers and black farm workers, especially since early 2000. The liberation war veterans and other government supporters, including some from urban areas who can no longer cope with the high cost of living on their meagre urban incomes, have occupied over 1,500 farms across the country. In the meantime, in the twelve months following the parliamentary general election of June 2000, the government designated some 3,000 commercial farms totalling about 5m. hectares for compulsory acquisition and resettlement by smallholder farmers under what it calls the 'fast-track' resettlement programme.

There is no doubt that the impacts of the 'fast-track' resettlement programme will be extensive and multi-dimensional. Detailed studies are needed on the emerging economic, social and environmental impacts of the programme. Its proponents argue that, apart from addressing the historical inequalities between Africans and Europeans, it will also bring about improved living standards and higher incomes for the settlers, while the intensification of land use will bring into production land that has until now been under-utilised.

On the other hand, critics of the programme have pointed to a number of negative impacts, among which are the following:

- There is a possibility that the country will not be able to feed itself, especially in the short term before the new settlers are sufficiently skilled and experienced producers. Because of the disruption to agricultural production caused by the farm invasions during the 2000-01 season, the country faces a substantial shortfall in its basic food requirements. The cereals shortfall to meet the country's needs to the next harvest in mid-2002 has been estimated as high as 500,000 tonnes of maize and 150,000 tonnes of wheat, which would cost as much as US$90m. to import.
- There has been drastic loss of Zimbabwe's foreign currency earnings due to the reduction in production of export crops such as tobacco and horticultural produce by commercial farmers, and the drop in tourist arrivals since early 2000. Agricultural exports and tourism are among the principal sources of foreign currency for the country. The violence associated with the 2000 general election and the land invasions since then has received wide publicity in the international media, and this has also contributed to the slump in tourist arrivals. Bed occupancy

rates in the hotel industry, including the major tourist resort of Victoria Falls, have declined to below 30 per cent since early 2000.

- Most international airlines have stopped flying into Zimbabwe. The loss of foreign currency has been most evident in the acute shortage and erratic supplies of fuel and imported raw materials for industry. As a result, several hundred factories have closed down and hotels have reduced their labour, thereby adding to the burgeoning unemployment.

- The 'fast-track' land reform programme has led to the loss of employment and the displacement of several thousands of farm workers and their families, many of whom are not being accommodated in the resettlement schemes. Some of these farm workers are second and third generation Zimbabweans whose parents originally came from neighbouring Mozambique, Malawi and Zambia. They have no historical links with the communal lands where they could seek to be accommodated. Their plight remains unresolved.

- Financial institutions have also been negatively affected because of their high exposure to the commercial agricultural sector. In order to reduce risk, banks and other lending institutions have refused to provide credit to farmers whose farms have been designated for resettlement by the government or are occupied by the invaders. This means that farmers who may still be on the land and willing to continue production are restricted by lack of seasonal finance.

- The outbreak of foot-and-mouth disease in August 2001, which resulted in a total ban on beef exports to the European Union and other regional markets, was attributed to the breakdown by the farm occupiers and poachers of game fences along the borders of Zimbabwe's national parks which are home to the wild buffalo which carries the virus, coupled with the uncontrolled movement of livestock into the parks.

- Finally, environmentalists, commercial farmers who have been practising game ranching, and the tourism industry are concerned about the environmental impacts of the resettlement programme, especially deforestation during land clearance and game poaching.

The recent occupations of European-owned land in Zimbabwe, with the tacit approval of the government and the ruling political party, also have regional implications. Following on the example set by the liberation war veterans and other landless people in Zimbabwe, similar land invasions have occurred in several other countries in the region, from Kenya in the north, through Malawi to Namibia and in South Africa. To date, governments in these different countries have responded in varying manner to the land occupations, but in general, they have tried to contain the situation while minimising the political cost of not being seen to support land reform and redistribution to the landless. According to Moyo (2000), variations in approach to land reform, both across countries and within countries in the region, relate to whether priority is given to redressing past injustices and the promotion of justice and equity, or to the narrower concerns of economic efficiency.

Conclusions

The chapter has examined the interaction between politics and economy and how that has created a social formation that is characterised by persistent spatial and social inequalities. The state in Zimbabwe, both before and after independence in 1980, has played a pivotal role in the evolution of these inequalities, and more recently, in efforts to redress them. Land reform and redistribution remains an unresolved and increasingly problematical issue with wide and over-arching implications for the socio-economic and political geography of both Zimbabwe and the southern African region.

By the late 1980s, the government's socialist policies had achieved a lot in terms of narrowing the gap between rural and urban areas and between rich and poor. However, the first decade after independence was characterised by a general deterioration in economic conditions that compelled the government to abandon many of its social policies and programmes. For a while in the 1990s, the government turned to the Bretton Woods institutions for balance of payment support as it endeavoured to restructure and liberalise the economy. Unfortunately, the economic reforms generally failed to address the problems, and economic decline has continued to this day, characterised by high unemployment, hyper-inflation, flight of capital and de-industrialisation, declining foreign currency earnings, deteriorating social services and widespread poverty in a society devastated by the HIV/AIDS epidemic.

References

Central Statistical Office and Macro International Inc. (1995), *Zimbabwe Demographic and Health Survey 1994*, Calverton, Maryland.

Central Statistical Office and Macro International Inc. (2000), *Zimbabwe Demographic and Health Survey 1999*, Calverton, Maryland.

Chenje, M., Sola, L. and Paleczny, D. (eds) (1998), *The State of Zimbabwe's Environment 1998*, Ministry of Mines, Environment and Tourism, Harare.

Drakakis-Smith, D., Bowyer-Bower, T. and Tevera, D.S. (1995), 'Urban Poverty and Urban Agriculture: an Overview of Linkages in Harare', *Habitat International*, vol. 19, pp. 183-93.

Fultz, E. and Pieris B. (1998), *The Social Protection of Migrant Workers in South Africa*, ILO/SAMAT, Harare.

Grizic, P.M. (1980), 'Water, the Vital Resource', *Zimbabwe Science News*, vol. 14, pp. 297-98.

Heath, R.A. (1992), 'Wildlife-based Tourism in Zimbabwe: an Outline of its Development and Future Policy Options', *Geographical Journal of Zimbabwe*, no. 23, pp. 59-78.

Herbst, J. (1990), *State Politics in Zimbabwe*, University of Zimbabwe Publications, Harare.

Mazvimavi, D. (1998), 'Water Availability and Utilisation in Zimbabwe', *Geographical Journal of Zimbabwe*, no. 29, pp. 23-36.

McDonald, D.A. (ed.) (2000), *On Borders: Perspectives on International Migration in Southern Africa*, Southern African Migration Project, St. Martin's Press, New York.

Mlambo, A.S. (1997), *The Economic Structural Adjustment Programme: The Case of Zimbabwe, 1990-1995*, University of Zimbabwe Publications, Harare.

Moyo, S. (2000), 'The Land Question and Land Reform in Southern Africa', in D. Tevera and S. Moyo (eds), *Environmental Security in Southern Africa*, SAPES Books, Harare, pp. 53-74.

Mudimu, G.D. (1996), 'Urban Agricultural Activities and Women's Strategies in Sustaining Family Livelihoods in Harare, Zimbabwe', *Singapore Journal of Tropical Geography*, vol. 17, pp. 179-94.

Ngara, T. and Rukobo, A. (1999), *Environmental Impacts of the 1991/92 Drought in Zimbabwe: an extreme event*, Radix Consultants (Pvt) Ltd, Harare.

Potts, D. and Mutambirwa, C. (1990), 'Rural-urban Linkages in Contemporary Harare: why Migrants need their Land', *Journal of Southern African Studies*, vol. 16, pp. 677-98.

Rakodi, C. (1994), 'Urban Poverty in Zimbabwe: Post-independence Efforts, Household Strategies and the Short-term Impact of Structural Adjustment', *Journal of International Development*, vol. 6, pp. 660-63.

Republic of Zimbabwe (1983), *Annual Report of the Secretary for Education for the year ended 31st December 1981*, Harare.

Republic of Zimbabwe (1991), *Annual Report of the Secretary for Education and Culture for the year ended 31st December 1990*, Harare.

Rukuni, M. and Eicher, C.K. (eds) (1994), *Zimbabwe's Agricultural Revolution*, University of Zimbabwe Publications, Harare.

SADC-ELMS (1997), *The Fundamentals of Natural Resources Policy Analysis for the SADC Region*, Southern African Development Community-Land Management Sector (SADC-ELMS), Maseru, Lesotho.

Swain, A. and Stalgren, P. (2000), 'Managing the Zambezi: the Need to Build Water Institutions', in D. Tevera and S. Moyo (eds), *Environmental Security in Southern Africa*, SAPES Books, Harare, pp. 119-38.

Taylor, R.D. (1988), 'The Indigenous Resources of the Zambezi Valley: an Overview', *Zimbabwe Science News*, vol. 22, nos. 1/2, pp. 5-8.

Tevera, D.S. (1994), 'Population, the Environment and Resources in Zimbabwe', in E. Kalipeni (ed), *Population Growth and Environmental Degradation in Southern Africa*, Lynne Reinner Publishers, London, pp. 39-59.

Tevera, D.S. (1995), 'The Medicine that might Kill the Patient: Structural Adjustment and Urban Poverty in Zimbabwe', in D. Simon, W. van Spengen, C. Dixon and A. Narman (eds), *Structurally Adjusted Africa: Poverty, Debt and Basic Needs*, Pluto Press, London, pp. 79-90.

Tevera, D.S. (1997), 'Structural Adjustment and Health Care in Zimbabwe', in E. Kalipeni and P. Thiuri (eds), *Cultural and Demographic Aspects of Health Care in Contemporary sub-Saharan Africa*, IASS, Langley Park, pp. 229-44.

Tevera, D.S. and Dahl, J. (2000), 'Return Migration in Zimbabwe: an Escalating Trend during the 1990s?', seminar paper presented in the Department of Human and Economic Geography, Gothenburg University, Gothenburg, Sweden, 23 November.

Whitlow, R.J. (1988), *Deforestation in Zimbabwe: A Geographical Study*, Natural Resources Board, Harare.

Whitlow, R. and Zinyama, L.M. (1988), 'Up Hill and Down Vale: Farming and Settlement Patterns in Zimunya Communal Land', *Geographical Journal of Zimbabwe*, no. 19, pp. 29-45.

World Bank (1989), *Sub-Saharan Africa: From Crisis to Sustainable Growth*, Washington DC.

Zinyama, L. and Whitlow, R. (1986), 'Changing Patterns of Population Distribution in Zimbabwe', *GeoJournal*, vol. 13, pp. 365-84.

Zinyama, L.M. (1986), 'Agricultural Development Policies in the African Farming Areas of Zimbabwe', *Geography*, vol. 71, pp. 105-14.

Zinyama, L.M. (1989), 'Multinational Disinvestment: Localisation or Socialist Transformation in Zimbabwe's Manufacturing Sector?', *Area*, vol. 21, pp. 229-35.

Zinyama, L.M. (1990), 'International Migrations to and from Zimbabwe and the Influence of Political Changes on Population Movements, 1965-1987', *International Migration Review*, vol.24, pp. 748-67.

Zinyama, L.M. (1995), 'Dams, People and Environmental Management in Zimbabwe', *Geographical Journal of Zimbabwe*, no. 26, pp. 17-29.

Zinyama, L.M. and Zanamwe, L. (1997), 'Who is Going down South? Gender Differences in Cross-border Movements of Zimbabweans to South Africa', *Geographical Journal of Zimbabwe*, no. 29, pp. 37-48.

14 Mozambique: Development, Inequality and the New Market Geography

GRAHAM HARRISON

Introduction

Mozambique has endured a more troubled history than any other country in the region, with the obvious exception of Angola. In a profound historical sense, the *mere existence* of Mozambique has always been problematic: as much an idea as a substantive social entity; a struggle of state power over an unlikely geography; a weak economy pervaded by global and regional powers with their own motivations. This chapter will explore these themes, with a bias towards Mozambique's contemporary conjuncture, framed by the region as well as its domestic politics. The rest of this introduction will briefly outline Mozambique's key features.

Current perceptions of Mozambique are framed by the country's poverty. The two images of the country that have momentarily seized global attention are the floods of 2000 and the high levels of debt that led to Highly Indebted Poor Country status. Mozambique has a GNP per capita of about $210, a life expectancy of 45 years and a high rate of HIV seroprevalence (World Bank 2000, pp. 6, 322). Interestingly, Mozambique is also one of a small group of African states that has performed relatively well under World Bank and IMF-sponsored structural adjustment programmes (SAP). Its real GDP growth in 1993 was 19.3 per cent, the highest in the world, but growth rates have stayed around 5 per cent since then. This apparent paradox, of natural disaster and poverty against market liberalisation and growth, brings us to the core concern of this chapter: to investigate Mozambique's contemporary socio-economic situation by evaluating the developmental prospects of what is termed here Mozambique's *new market geography*.

Mozambique was created in 1910 when its western boundaries were agreed between Portugal and Britain. It remained a Portuguese colony until 1974, when it underwent a year's transitional government followed by independence under the single-party regime of the *Frente da Libertação de Moçambique*, Frelimo. Frelimo's consolidation of power led Mozambique to develop one of the boldest socialist experiments in the periphery, encapsulated in the ostentatious Decade of Victory Over Underdevelopment (1980-1990). During this period, an insurgent force known as Renamo (*Resistência Nacional de Moçambique*) became a serious national security concern as it moved into southern Mozambique armed and trained by apartheid South Africa. The war fomented by Renamo was a key reason for the collapse of Frelimo's state-planned modernisation, although it was certainly not the sole reason. The failure of Frelimo's vision led to a piecemeal process of reform which eventually consolidated a fairly radical liberalisation of the economy, especially after 1987 when Mozambique signed its first structural adjustment programme, PRE, in 1987. Economic change was succeeded by political change: in 1989 Renamo and Frelimo commenced negotiations that would lead to the Rome Peace Accords of 1992 and elections in 1994. Frelimo has remained in government, and the peace has held.

Having made a brief review of Mozambique's basic characteristics, let us re-state our critical interest: how has Mozambique's new market geography affected its economy and development? As always, the first step in understanding the present is to acknowledge the influence of the past.

The Mozambican State and Economy in Historical Perspective

Portuguese Colonialism

Despite Portuguese claims to a five hundred-year empire in sub-Saharan Africa, the colonial state only established the bare bones of a national political presence in Mozambique after the First World War. Much of Mozambique's early colonial history could be characterised as 'rentier colonialism', in which Mozambique's territory was leased to non-Portuguese interests. It was only after 1926 (when António Salazar came to power in Portugal) that the colonial project in Mozambique attempted seriously to forge a national economy, oriented towards the Portuguese metropole.

Forced labour was widespread until 1961, and carried on in covert forms up until independence. Refusal to comply with colonial labour regimes would lead to beatings, and a period of unpaid labour in colonially-owned plantations (Isaacman, 1996). Mozambique was subjected to what has been termed 'shopkeeper colonialism': Portugal was not as economically powerful as France or Britain, and as a result much of the influx of Portuguese came in the form of small shopkeepers and farmers. These traders and farmers produced crops for Mozambique's cities and relied on the exploitation of Mozambican peasants and privileges from the colonial state (Mackintosh, 1987). From the 1960s, Mozambique experienced a rapid build-up in the industrial and construction sector, leading to the creation of a Mozambican working class, especially in the ports (Wield 1983, p. 76 *et seq.*).

The legacy of colonialism was a country divided into three broad regions with different economic structures (Wuyts, 1980). To the south of the Save River, a labour reserve was established in the early 1900s, and consolidated in the 1960s when around 100,000 Mozambican men migrated to the mines and plantations of South Africa each year (First, 1983). In the centre, plantations drew peasant labour into the colonial economy, and men also migrated to Rhodesia in search of wage labour. In the northern third of the country, peasants were locked into a regime of forced cotton cultivation. This economic geography created a national economy which was poorly integrated and dependent on close relations with neighbouring states.

Regional Configuration

Let us pursue the issue of Mozambique's role in the southern African region a little further. The importance of this aspect of Mozambique's economic geography derives from the fact that Mozambique was dependent on its relations with South Africa (Alden, 1992). There are three key characteristics that define this regional dependence.

- Migrant labour. The Witwatersrand Native Labour Authority (WENELA) recruited thousands of Mozambican men to work in South Africa's mines. In the late 1960s and early 1970s, recruitment peaked at 110,000. The 'real' level of migration would be even higher because there was a vibrant illicit migration taking place concurrently. This created a rural peasant economy which was fully integrated with a system of migration. Remittances, new forms of family relations, new career pathways and forms of accumulation, and indeed new forms of culture emerged around the migration experience (First, 1983; Harries, 1994; van den Berg, 1987).

- External Income. WENELA acted as the monopoly recruiter of Mozambican labour for the South African mines. It also paid the miners' wages. Importantly, wages were paid in two ways, about half to the miner at the end of contract and about half to the Mozambican state in gold, at a fixed rate. The latter was set below the free market price, allowing the Mozambican state to sell the gold at a profit, which it could add to the coffers after paying miners in Portuguese escudos on their return. In 1978, after the IMF abolished all controls on gold prices, this source of income ended. Thus, not only did the peasant societies of southern Mozambique come to depend on mine labour, so did the colonial coffers.

- Serving as an entrepôt for much of the growing mineral-industrial development of the Pretoria-Witwatersrand-Vereeniging (PWV) region, Maputo earned revenues as a port and rail terminal for much South African trade with the rest of the world. The same entrepôt function breathed life into Mozambique's second city, Beira, which served the Rhodesian economy. Both Rhodesia and South Africa also supplied thousands of white tourists who found cheap seafood, beautiful beaches, and opportunities for sex tourism in Mozambique.

In sum, colonialism created a state in Mozambique with official jurisdiction over an established territory. But the socio-economic content of that political jurisdiction was riven with contradictions: inasmuch as a national economy existed in Mozambique, it did so substantially by virtue of its dependent relations with South Africa and to a lesser extent Rhodesia. This long coastal state served as an entrepôt, but few things tied together the consciousness or economies of the people of Cabo Delgado in the north to Gaza province in the south (Figure 14.1). This difficult legacy profoundly shaped the prospects for any post-colonial government, which is our next concern.

Frelimo's Socialism

The *Frente da Libertação de Moçambique* (Frelimo), created in 1962, fired the first shots of the anti-colonial war in 1964. Pursuing a war of attrition against an increasingly dissatisfied colonial army, Frelimo managed to gain a presence in the northerly third of the country before bringing the Portuguese to the negotiating table. In 1974, the Lusaka Accords were signed and, after a year's transitional government, Mozambique gained independence in 1975.

After a period of instability, generated by the massive perturbations of a Portuguese exodus, Frelimo established a Marxist-Leninist political system (Egerö, 1990) accompanied by state-centred development planning. Hailed as the beginning of a rapid process of modernisation which would bring Mozambique out of the Third World by 1980, Frelimo's socialism contained the following key features:

Figure 14.1 Mozambique

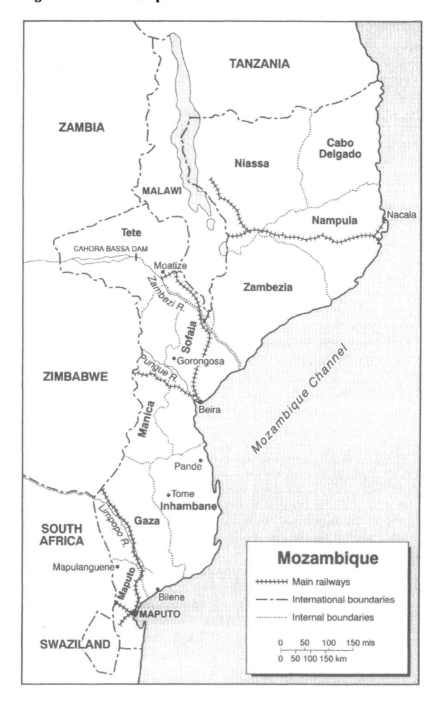

- The creation of large state-owned companies. Initially an *ad hoc* reaction to Portuguese exit and sabotage, Frelimo nationalised private property. Subsequently, companies were created not as a means of crisis management, but to ensure the success of input-based national planning.

- The creation of state-owned agro-industrial plantations. Rural areas that enjoyed well-established infrastructure were integrated into large state-owned irrigated and mechanised farms, producing crops for the urban population (Raikes, 1984). As with much of Frelimo's socialist development planning, financial and technical support from the Soviet bloc was pivotal.

- Villagisation. Peasants were to be concentrated in communal villages, ordered around the Frelimo party-state, working on state-owned plantations or collective farms. There were 5m. in communal villages by 1990 (Hall and Young, 1997, p. 96).

- State-controlled trade and exchange. Frelimo created trading institutions, and set prices for most goods bought and sold. Permission was required to trade across provinces. Frelimo controlled the exchange rate, allowing it to be substantially over-valued.

These four features of Frelimo's economic model were accompanied by a changing regional political economy. Frelimo supported sanctions against the Smith regime in Rhodesia, and allowed ZANU guerrillas to use central Mozambique as a rear base. It also declared its support for the ANC, leading to a brief occupation of southern Mozambique by the South African Defence Force and the cutting of the number of Mozambican migrants from 115,000 in 1975 to 40,000 in 1976. Thus, Mozambique's dependent insertion into the region was replaced by a more antagonistic stand-off, militarised as a result of the general regional conflict between racial privilege and African nationalism.

Frelimo inherited the legacy of a nation-state born out of the contradictions of Portuguese colonialism. Frelimo's virile nationalism could not resolve the fundamental weaknesses left by the colonial period. Regional support had to be replaced by other external sources. Consequently, from 1977 until the mid-1980s, Mozambique came to reply heavily on Soviet and Soviet-aligned aid, loans, and technical expertise (Hall and Young 1997, p. 138 *et seq.*). Also from the late 1970s, the non-aligned states implemented large aid programmes, especially in the agricultural sector.

Frelimo's plans contained many contradictions. Development was technocratic and bureaucratic in conception and authoritarian in execution.

Peasants often resented their movement into communal villages, and the state-owned farms rarely turned a profit (Hermele, 1988; Casal, 1988; Geffray, 1991; Cardoso, 1993; Harrison, 2000). State-owned industries ran up large debts, produced inefficiently, and did not renew capital stock. These failings led to a gradual process of liberalisation from the Fourth Party Congress in 1983. However, the success or otherwise of Frelimo's audacious plans, or their reform, was pre-empted by the rise of the *Resistência Nacional de Moçambique* (Renamo).

Renamo and Regional Destabilisation

In the late 1970s, Renamo was created in Rhodesia, initially to attack guerrilla ZANU bases in western Mozambique, but also to undermine Frelimo's presence in the unstable border areas between the two countries. After 1980, Renamo was trained, funded, and instructed by the South African military apparatus. South Africa's strategic goal was to ensure that Frelimo suffered for its vocal opposition to apartheid and its aspirations towards Soviet-aligned socialism. South Africa pursued a war against the region, differentiated in its measures and specific aims, known as Total Strategy (Hanlon, 1986; Kibble and Bush, 1986; Fauvet, 1984). In Mozambique, Renamo destabilised Frelimo's development plans by destroying evidence of investment in schools, health posts, and infrastructure. It also terrorised the rural population in order to undermine Frelimo's power structures. It subjected Frelimo cadres (including teachers as well as village presidents) to public torture and murder (Minter, 1989; Cammack, 1988).

By 1984, Renamo had a foothold in every province of Mozambique. By the end of the 1980s, the entire Mozambican economy (such as it was) had been rent apart by war. Economic interchange was literally militarised: trade corridors emerged between major cities and between Mozambique and its neighbours, patrolled by the Mozambique Armed Defence Forces, with support from the Zimbabwean and Tanzanian military. Cities were rendered islands of party-state power within unstable seas of insurgency and counter-insurgency. Peasants were prey for both protagonists – sources of new troops, food, intelligence, or resources in themselves, to be forcibly removed from the enemy's influence. The only group to profit were those who could make cash from chaos by exploiting lawlessness or contested jurisdiction to traffic and trade, often across national borders. This group included ivory traders within Renamo and war material traders within the military.

Economic degradation constituted one of a raft of factors that conspired to bring the civil war to an end. Drought in the early 1990s meant that both Renamo and Frelimo had difficulty feeding their troops and ensuring their loyalty, let alone motivation to fight. Concurrently, the South African state began a complex transition to end a constitutionalised racism in South Africa. This meant a tailing off of South African support for Renamo during the late 1980s. As Renamo lost its means to wage war, Frelimo faced equally pressing circumstances. The war economy was unsustainable: in 1987, 50 per cent of the budget went to the military. Mozambique's debt burden to the West increased massively in the late 1980s, during which time *perestroika* and *glasnost* moved Western attention (and indulgence) away from Africa and towards the Eastern bloc, which itself had reduced its interest in supporting Mozambique.

The outcome of this conjuncture was a convoluted series of ceasefire and peace negotiations, commenced in Kenya and completed in Rome with the General Peace Accords of 1992 (Vines, 1991). The peace held and ushered in Mozambique's first multi-party elections in 1994, to be followed by a second round in 1999.

This short history brings us to the current period of Mozambique's new market geography. Before we take this period in more detail, let us draw some lessons from the above history to carry over into the following sections:

- Mozambique's post-colonial history was one of decline: a short period of 'authoritarian modernisation' was succeeded by a general decline – substantially a result of the war against Renamo – that has left the country failing to reach the levels of economic output of the early 1980s.
- There has been a continuity of power in Mozambique. Frelimo remains the incumbent, and many of the 'old guard' remain key figures in the cabinet. The core area of that state, and its social origins remains embedded in the southern region of Mozambique. Some would trace this geographical concentration to the colonial period, when the capital of the colony was transferred from the Ilha de Moçambique to Maputo in 1901 (Cahen, 1993).
- The war had created a regionalised illicit economy and a kind of brutal 'grassroots' capitalism forged in the social spaces that the state failed to control (Chingono, 1996).

The Aftermath: Reconstruction and Regional Reconfiguration

In order to understand the general context of change that frames Mozambique's current new market geography, we need to make a brief review of Mozambique's experience with structural adjustment, the *Programa de Reabilitação Económica* (PRE) from 1987 to the present day. Mozambique joined the World Bank and International Monetary Fund (IMF) in 1984, in the early period of gradual liberalisation mentioned above. In 1987, Mozambique implemented a structural adjustment programme with World Bank funding. In keeping with structural adjustment elsewhere, PRE contained key generic elements:

- radical devaluation of the Mozambican currency, the metical;
- divestment of state-owned enterprises (SOE);
- removal of price controls and subsidies;
- measures to control money supply;
- liberalisation of foreign trade and investment;
- reduction of the budget deficit (from Hermele, 1988, Table 1).

The Frelimo government implemented a fair amount of the PRE, bearing in mind its declared historic origins in Marxism-Leninism, and the context of intense civil war: the metical's (MT) value to the dollar fell from MT525 to MT2,476 between 1988 and 1992 (EIU, 1993, 2nd Quarter: 3); subsidies and rations on consumer goods (which had been gradually removed from the early 1980s) were reduced or removed (O'Laughlin, 1996); a concerted programme of divestment was implemented throughout the 1990s (Cramer 2001); foreign exchange shops were legalised, effectively unifying official bank and free market rates for currencies; institutional and administrative reform reduced certain areas of public expenditure. But the PRE did not proceed smoothly, even if progress towards its objectives was made. There are three factors which explain the limits to progress and the instability of reform.

Firstly, the PRE was implemented during a war. Amazingly, neither the World Bank nor the IMF made any serious acknowledgement that the war affected Mozambique's economy. Rather, the war was factored out as an external condition which, it was assumed, would eventually be neutralised. The result of this was that the Mozambican masses suffered extremely high levels of poverty, deprivation, and dislocation whilst the PRE dictated that social spending be reduced as a key measure to reduce public expenditure and work towards a balanced current account. During the war, Renamo destroyed 31 per cent of the health care system in rural areas (Hermele, 1990, p. 6). Despite this, health expenditure was cut from $5 per person in 1981 to an average of $1 per person in 1988 (ibid., 1990, p. 27), as part of the PRE.

Secondly, the Frelimo state was destabilised by the PRE, as there were strong constituents both advocating and resisting free market reform

(Simpson, 1993). Reform was resisted not only by the 'old ideologues' of the Marxist Leninist period, but by those 'new entrepreneurs' who were making profits from the spoils that came from opaque links between public office and private business (Bowen, 1992). Support for the PRE emerged among the younger 'technocrats' within the government, and specifically within the Ministry of Finance. As a result of the political contests raging behind the scenes, Mozambique experienced a fair degree of 'policy slippage' whereby reforms were either partially implemented or evaded (World Bank, 1994).

Thirdly, the PRE provoked popular resistance in the urban areas. Real wages were falling, subsidies on staple goods were removed, fuel prices increased (making public transport more expensive), and retrenchment pushed larger numbers of people into the informal economy. Generally, the social costs of the PRE were substantial (Marshall, 1992). As a result, strikes and riots featured in Mozambique's political landscape in the early 1990s.

In 1990, the PRE was replaced by a second structural adjustment programme with a stronger social component (Abrahamsson and Nilsson, 1995, p. 112). In 1992, the Rome Peace Accords were signed, ushering in an end to war. Foreign investment picked up, especially from Portugal, South Africa and the United Kingdom: after levels of FDI at around $20m per year in the early 1990s, rates rose strongly in the latter half of the decade: $66m in 1994 and $445.5m in 1996. Simplified investment legislation was introduced in 1993. Peasant farmers returned to their land and a reasonable level of social stability, generating an increase in agricultural output. The World Bank and other donors ploughed substantial funds into Mozambique, allowing the government formally to balance its budget despite the fact that by 1990, 75 per cent of its budget came from external sources. In 1993, GDP increased by 19.3 per cent, and subsequently the economy has expanded by about 5 per cent per year. In essence, Mozambique has experienced economic recovery during the 1990s, at least in terms of macro-economic aggregates. One result of this has been the solidifying of policy reform, deepening from 'stroke of the pen' liberalisation to more complex areas of administrative and technical reform. The Frelimo state developed increasingly close relations with the World Bank, IMF, and the coterie of other donor/creditors that moved into Maputo. The most recent ratification of this changing state of affairs is Mozambique's winning of Highly Indebted Poor Country status (HIPC) in 1998, which reduces Mozambique's indebtedness from about 400 per cent of export earnings to about 200 per cent, keeping the debt-service ratio at a 'sustainable' 20 per cent. The real financial benefits to Mozambique have been debated, but HIPC also opens the way to close and supportive relations with bilateral donors, who perceive HIPC status as a 'stamp of approval' by the international financial institutions.

The progress of the PRE provides a strong context within which to understand more specific features of spatial and economic change. The PRE has liberalised the Mozambican economy, but not in uncontroversial fashion. Reforms appeared extremely shaky for the first five years, to be succeeded by a period of economic recovery. As a result of this turnaround in the PRE's fortunes, the free market agenda has been 'locked in' to Mozambique's policy debate. But this is not to say that Mozambique's new market geography constitutes a kind of spatial harmonisation of a formerly war-ravaged and statist economy. As we shall see, the real processes of economic growth have been accompanied by equally real contradictions, continuities from previous ages, and limits in the prospects for recovery.

Features of the New Market Geography

The purpose of this section is to trace the 'sinews' of Mozambique's new market geography, forged during peacetime and the PRE. Because the focus is very much towards present-day change in Mozambique, we are defining tendencies as much as robust structures. The aim is to outline a convincing picture of Mozambique's transition and its prospects, and to raise salient issues that will help us evaluate the latter.

Tourism

Mozambique occupies a long stretch of Indian Ocean coast, with islands of exceptional natural beauty or cultural heritage. As Mozambique has opened to foreign investment, companies have become interested in investing in luxury tourism, serving Western markets. The most prominent symbol of this opening was the refurbishment of the Polana hotel in Maputo in 1992 by its South African owners. Most (in)famously, James Blanchard III, an American millionaire who had close links with Renamo during the war, gained permission to create a luxury 'eco-tourism' park, covering 200,000 hectares in southern Mozambique, although the likelihood of this project being realised is rather doubtful. More modest and realistic projects have been carried out, however. Large luxury hotel chains have rehabilitated Maputo's hotels, and constructed new ones. Small plane companies have sprung up to take tourists to and from islands and into game reserves. Smaller companies serve tourists by providing fishing expeditions. Much of the capital for new tourist investments has come from South Africa and the large international hotel chains.

The vast majority of investment in tourism has been focussed in the south, along the Costa do Sol, Inhambane, and the Islands facing this southern strip. A smaller rehabilitation of tourism has occurred in the extreme north: in Cabo Delgado's provincial capital, Pemba, a coastal complex of luxury cabins has been rehabilitated and expanded and, during this author's last visit (1998), there were plans for a new tourist complex nearby.

Industry

Mozambique's industrial sector has been through turbulent times during liberalisation. State-owned enterprises were often inefficient and drained the public purse. However, the fortunes of industry under privatisation have been variable: industrial production declined from 1990 to 1995 (Tibana, 1995, p. 1) and recovered thereafter. Privatisation has been only minimally monitored and not subject to any strategic 'development' framework by the government (Cramer, 2001).

Mozambique has created a series of Industrial Free Zones (essentially Export Processing Zones) on the coast which encourage export-based manufacture. Basic consumer goods industry has recovered since the end of the war, but is limited by Mozambique's small domestic market. Perhaps the greatest success story here is the bottled drinks industry. The rapid growth of the cement industry reflects the amount of rebuilding and new construction which is currently taking place. A more problematic example is cashew processing (Cramer, 1999).

Mining and Energy

Mozambique drains much of the region's large river systems (Figure 14.1). It also has unproved offshore and onshore oil and gas, as well as gold and minerals such as tantalite and graphite. Foreign companies have committed substantial sums of investment in mineral extraction. The largest single FDI project, agreed but not yet implemented, is the Maputo Steel Project which will process ore from South Africa for global export. This involves $2bn. of foreign direct investment (FDI), mainly from South Africa and the USA. Enron, an American company, is investing in the Pande gas fields: the associated power plant, to be located in Maputo, will supply the Maputo Steel Project, as well as the South African market. The Mozal plant in southern Mozambique will process aluminium for export to South Africa, and is projected to bring $1.4bn. of FDI into the country – the second largest single foreign investment in Mozambique's history. Lonrho is prospecting for oil in the north. In sum, the energy/minerals sector has attracted a series of so-called mega projects into Mozambique.

Transport

Mozambique has received large amounts of soft loan money from the World Bank and others to reconstruct its devastated road and rail system. The construction projects (ROCS I and II) have concentrated on national highways. The Ports at Maputo and Beira (serving Zimbabwe) have both experienced a healthy increase in traffic. The most innovative development in Mozambique's transport system has been the creation of the Maputo Development Corridor, running to Witbank. A toll road is being constructed on a build-operate-transfer contract with the TRAC consortium (involving French, South African, Portuguese and Mozambican capital) which will, it is hoped, integrate South Africa's mineral-industrial heartland with Maputo as a port. The model is that the line of road will produce a 'ribbon' of development as trade and industry is attracted to the market produced by transport and the low costs of locating near to the high-speed highway (Rogerson, 2001). In other parts of the country, smaller projects of road rehabilitation proceed in a more cautious fashion.

Agriculture

Mozambique's agricultural recovery is pivotal to the country's sustainable development (Ferraz and Munslow, 2000). Liberalisation has brought both costs and benefits to peasant farmers. Peasants are no longer subject to the onerous regulations of Frelimo's 'socialism'. Crop prices have increased during liberalisation, providing a greater set of incentives to trade (or not to sell crops on parallel markets). Obviously, the end of war has ushered in a great improvement to peasants' lives with strong repercussions on the efficiency of production (Harrison, 2000). The Frelimo government has not developed a clear programme for agricultural development until very recently (Woodhouse, 1997); the PROAGRI plan, only completed in 1999, is based on basic social provision, infrastructural development, and strong donor support. The hub of agricultural recovery has come from the river basins, especially where a class of relatively wealthy farmers has persisted or developed. It is less clear that the peasant farmers in rain-fed areas, more remote from market structures, are participating fully in agricultural recovery.

The Illicit/Informal Economy

Mozambique's economic transformation has been only partly visible. Much of Mozambique's changing class relations and patterns of trade and accumulation traverse the boundaries of legality (Harrison 1999a). In the first place, peasants and wage labourers routinely engage in forms of informal economic activity, producing artisanal goods, small snacks, services, and trading. There are no statistics to evaluate the scope and nature of this economy, or its connections with formal trade and production.

At another level, liberalisation has provided the scope for well-placed members of the elite to indulge in corrupt business practices (Wuyts 1996: 728; Harrison 1999b). Privatisation generated a rather opaque scramble for cut-price former state property. There has also been a growth in drugs trafficking, especially in the manufactured drug Mandrax. In 1994, about one tonne of Mandrax was being detained by customs in Maputo *every week* on its journey from India to South Africa (EIU 1994, 2nd Quarter: 15). Large amounts of trade evade customs duties by employing bribes or smuggling: in 1990, official exports from Mozambique tallied $126m.; goods imported from Mozambique in trading partner states amounted to $363m. (EIU 1996, 1st Quarter: 12). The weakness of border controls allows the illicit import and export of a range of commodities. In northern Mozambique, illicit gold mining has created a substantial plunder economy, with estimates of 13 tonnes of gold being smuggled out of Mozambique from Niassa province alone between 1990 and 1994 (EIU 1994, 3rd Quarter: 14).

Evaluating Mozambique's New Market Geography

The accepted motif to characterise contemporary Mozambique has been one of liberalisation, peace, and recovery. Recovery has been encouraging in many respects: output in all sectors of the economy has improved; Mozambique has managed to attract a substantial amount of FDI; external donors and creditors remain committed to Mozambique. However, there are some critical points that need to be raised to calibrate any assessment of the new market geography.

A Geography of Concentration

Much of Mozambique's recovery is concentrated in the south. Maputo city accounted for 34 per cent of the entire national GDP in 1998 (Alden, 2001,

p. 120). Seventy per cent of the population live in absolute poverty (Herbold Green and Mavie, 1994, p. 78). Many of the large tourist and transport projects are located in the south. It is unclear whether this concentration will generate 'trickle down' or 'spread' effects for the rest of the economy. Mineral investment, which has brought in the most FDI, is still subject to critiques about enclave economics. Relatedly, much of Mozambique's current transition is based in a re-integration with the South African economy, very much the dominant force in the region. Mozambique's southerly provinces constitute a market for exports (Alden, 2001, p. 86) (South African frozen chickens have priced out chicken domestically-produced by co-operative associations for example), and Maputo provides a potentially cheaper export outlet compared with Richards Bay. The issue here is whether Mozambique's prospects are helped or hindered by its reintegration with South Africa, as a subordinate and dependent partner. A controversial lens through which to view Mozambique's relations with South Africa is the Mosagrius project in which subsidised white South African farmers have located in the northerly Niassa province in the expectation that they will be able to generate a new source of surplus crop output. Results so far have been modest.

Transformation or Consolidation?

Many of the features of the new market geography evoke memories of older structures forged out of the colonial period. Not only is Mozambique resembling the 'service economy' of South Africa that it used to be in the late colonial period, but investment is concentrated in the south and in the other centres of the colonial economy. Infrastructural development has been based in similar areas, most obviously with the rehabilitation of the colonially-constructed Cahora Bassa dam. Mozambique's tourist centres recall the colonial eras of whites coming for prawns, white sands and sexual recreation: prostitution has increased and Maputo now has a downtown striptease nightclub. Perhaps the socialist critique of 'decadent' capitalism was not entirely misplaced ...

Most importantly, questions have been raised concerning the sustainability of recovery (Arndt et al., 2000): to what extent is recovery a one-off response to peace and liberalisation and to what extent will it engender a more profound structural transformation?

Markets and Differentiation

The market might have compelled higher rates of output or growth, but it has also led to social differentiation. Mozambique's proportion of absolute poor has not fallen significantly. The urban geography of the cities is one of ostentatious reconstruction and shanty town poverty. There is a racialised politics emerging from this, as certain parts of Maputo seem to be the preserve of European and South African tourists, consultants, and businesspeople who have minimal contact with the rest of Mozambique. The close juxtaposition of extreme wealth and poverty (in fact enrichment and impoverishment) generates social tensions, manifest in occasional riots, protests, violent crime, and the profusion of security companies which create a siege urban geography not dissimilar to the wealthy suburbs of South Africa. Serious questions have been raised about the way the PRE has intensified social differentiation (Oxfam, 1997), especially its paramount concern with fiscal austerity in a context of massive social deprivation: in 1997, debt repayment per person in Mozambique was $12 and expenditure on health was $5 per person.

A New Dependency?

In 1995, the debate on the budget in the elected Assembly of the Republic had to be delayed until the Paris Club of creditors had approved the plans. About two-thirds of the Mozambican budget typically comes from external sources and Mozambique became the most aid-dependent country in the world in the mid-1990s (Abrahamsson and Nilsson, 1995, p. 132). In spite of HIPC status, Mozambique remains trapped in a cycle of indebtedness and constant negotiation with donors/creditors. Mozambique's dependency has rendered purposeful public action increasingly problematic as the state is 'hollowed out' and/or divided up between different external donors (Wuyts, 1996). The Mozambican government's vulnerability to external agencies renders it supine to international investors, allowing companies to operate with little regulation of their activities, and even less revenue generation for the state from their activities. FDI reveals substantial continuities with the late colonial period: Portugal has recently become the largest investor in Mozambique, followed by South Africa, and then the UK, although gross levels of FDI reveal the UK as the most significant source of FDI overall.

In sum, Mozambique's new market geography suggests that its stability and developmental capacity are yet to be proven, even if Mozambique has undergone a form of recovery. It is worth bearing in mind that this country case study is extreme but not exceptional when it is placed in the regional context. Similarities and parallels exist over Mozambique's borders.

References

Abrahamsson, H. and A. Nilsson (1995), *Mozambique, the Troubled Transition: from Socialist Construction to Free Market Capitalism*, Zed Books, London.

Alden, C. (1992), 'Mozambique: an Abiding Dependency' in L. Benjamin and C. Gregory (eds), *Southern Africa at the Crossroads*, Justified Press, Johannesburg, pp. 74-102.

Alden, C. (2001), *Mozambique and the Construction of a New African State*, Palgrave Houndmills.

Arndt, C., Tarp Jensen, H. and Tarp, F. (2000), 'Stabilization and Structural Adjustment in Mozambique: an Appraisal', *Journal of International Development*, vol. 12, pp. 299-323.

Bowen, M. (1992), 'Beyond Reform: Adjustment and Political Power in Contemporary Mozambique', *Journal of Modern African Studies*, vol. 30, pp. 255-81.

Cahen, M. (1993), 'Check on Socialism in Mozambique: What Check? What Socialism?', *Review of African Political Economy*, vol. 57, pp. 46-59.

Cammack, D. (1988), 'The "Human Face" of Destabilization: the War in Mozambique', *Review of African Political Economy*, vol. 40, pp. 65-75.

Cardoso, F. (1993), *Gestão e Desenvolvimento. Moçambique no Contexto da África Sub sahariana*, Fim do Século, Oporto.

Casal, A. (1988), 'A Crise da Produção Familiar e as Aldeias Comunais em Moçambique', *Revista Internacional de Estudos Africanos*, vols. 8/9, pp. 157-91.

Chingono, M. F. (1996), *The State, Violence and Development: the Political Economy of War in Mozambique, 1975-1992*, Avebury, Aldershot, England.

Cramer, C. (1999), 'Can Africa Industrialize by Processing Primary Commodities? The Case of Mozambican Cashew Nuts', *World Development*, vol. 27, pp. 1247-66.

Cramer, C. (2001), 'Privatisation and Adjustment in Mozambique: a "Hospital Pass"?', *Journal of Southern African Studies*, vol. 27, pp. 79-105.

Economist Intelligence Unit (EIU), *Country Reports: Mozambique and Malawi*, EIU, London, various issues.

Egerö, B. (1990), *Mozambique, a Dream Undone: the Political Economy of Democracy, 1975-84*, Nordiska Afrikainstitutet, Uppsala.

Fauvet, P. (1984), 'Roots of the Counter-Revolution: the MNR', *Review of African Political Economy*, vol. 29, pp. 108-21.

Ferraz, B. and Munslow, B. (eds) (2000), *Sustainable Development in Mozambique*, James Currey, Oxford.

First, R. (1983), *Black Gold: the Mozambican Miner, Proletarian and Peasant*, Harvester Press, Brighton, Sussex.

Geffray, C. (1991), *A Causa das Armas: Antropologia da Guerra em Moçambique*, Afrontamento, Oporto.

Hall, M. and Young, T. (1997), *Confronting Leviathan: Mozambique since Independence*, C. Hurst, London.

Hanlon, J. (1986), *Beggar Your Neighbours: Apartheid Power in Southern Africa*, James Currey, London.

Harries, P. (1994), *Work, Culture, and Identity: Migrant Labourers in Mozambique and South Africa, c.1860-1910*, Heinemann, Portsmouth, New Hampshire.

Harrison, G. (1999a), 'Corruption, Development Theory, and the Boundaries of Social Change', *Contemporary Politics*, vol. 5, pp. 207-20.

Harrison, G. (1999b), 'Corruption as "Boundary Politics": the State, Democratization, and Mozambique's Unstable Liberalization', *Third World Quarterly*, vol. 20, pp. 537-51.

Harrison, G. (2000), *Grassroots Governance: The Politics of Rural Democratization in Mozambique*, Edwin Mellen, Lampeter.

Herbold Green, R. and Mavie, M. (1994), 'From Survival to Livelihood in Mozambique', *IDS Bulletin*, vol. 25, pp. 77-84.

Hermele, K. (1988), *Land Struggles and Social Differentiation in Southern Mozambique: a case study of Chokwe, Limpopo, 1950-1987*, Scandinavian Institute of African Studies, Uppsala.

Hermele, K. (1990), *Mozambican Crossroads: Economic and Politics in the Era of Structural Adjustment*, Chr. Michelsen, Bergen, Report 3.

Isaacman, A.F. (1996), *Cotton is the Mother of Poverty: Peasants, Work, and Rural Struggle in Colonial Mozambique, 1938-1961*, James Currey, London.

Kibble, S. and Bush, R. (1986), 'Reform of Apartheid and Continued Destabilisation in Southern Africa', *Journal of Modern African Studies*, vol. 24, pp. 203-27.

Mackintosh, M. (1987), 'Agricultural Marketing and Socialist Accumulation: a case study of Maize Marketing in Mozambique', *Journal of Peasant Studies*, vol. 14, pp. 243-67.

Marshall, J. (1992), *War Debt and Structural Adjustment in Mozambique: the Social Impact, North-South Institute*, Ottawa.

Minter, W. (1989), 'Inside Renamo', *Transformation*, vol. 10, pp. 1-28.

O'Laughlin, B. (1996), 'From Basic Needs to Safety Nets: the Rise and Fall of Urban Food Rationing in Mozambique', *European Journal of Development Research*, vol. 8, pp. 200-23.

Oxfam (1997), *Debt Relief for Mozambique: Investing in Peace*, Oxfam International, Washington DC.

Raikes, P. (1984), 'Food Policy and Production since Independence', *Review of African Political Economy*, vol. 29, pp. 95-107.

Rogerson, C. (2001), 'Road Construction and Small Enterprise Development: the Experience of the Maputo Development Corridor', *Development Southern Africa*, vol. 17, pp. 534-66.

Simpson, M. (1993), 'Foreign and Domestic Factors in the Transformation of Mozambique', *Journal of Modern African Studies*, vol. 31, pp. 309-37.

Tibana, R. (1995), 'Stabilization and Structural Adjustment in a Dual Transition: Mozambique in the 1990s', paper presented at workshop on 'Mozambique's Post-Electoral Challenges', Department of International Relations, London School of Economics.

Van den Berg, J. (1987), 'A Peasant Form of Production: Wage-Dependent Agriculture in Southern Mozambique', *Canadian Journal of African Studies*, vol. 21, pp. 375-89.

Vines, A. (1991), *Renamo: Terrorism in Mozambique*, Centre for Southern African Studies, University of York in association with James Currey, London.

Wield, D. (1983), 'Mozambique – Late Colonialism and Early Problems of Transition' in R. Murray and C. White (eds), *Revolutionary Socialist Development in the Third World*, Harvester Press, Brighton, Sussex.

Woodhouse, P. (1997), 'Virtue or Necessity? Pluralist Agricultural Technology Development in Mozambique', *Journal of International Development*, vol. 9, pp. 331-46.

World Bank (1994), *Adjustment in Africa: Reforms Results and the Road Ahead*, World Bank, Washington.

World Bank (2000), *African Development Indicators*, World Bank, Washington.

Wuyts, M. (1980), 'Economia Política do Colonialismo em Moçambique', *Estudos Moçambicanos*, vol. 1, pp. 9-22.

Wuyts, M. (1996), 'Foreign Aid, Structural Adjustment, and Public Management: the Mozambican Experience', *Development and Change*, vol. 27, pp. 717-49.

Part 3
Southern Africa and Beyond

15 Regional Economic Integration

RICHARD GIBB

Introduction

Trade and the creation of trading blocs has been, and continues to be, at the very centre of attempts to promote intra-African co-operation (Harvey, 1997). Throughout southern Africa the most commonly cited justification for regional economic integration is the small size of the region's individual economies. By joining together states are perceived to be in a better position to exploit larger-scale economies and, in theory, to restructure the regional economy in a way that benefits the production base of the region as a whole. In 2000 the Executive President of the Southern African Development Community (SADC), Dr Prega Ramsamy, argued for greater regional integration on the grounds that the 'individual economies of most member states are not internationally viable and competitive, as they are not in a position to enjoy the required economies of scale and to effectively deal with constraints to international competitiveness' (Ramsamy, 2000, p. 3).

From a global viewpoint there is little doubt that southern Africa, and for that matter Africa as a whole, represents a small and peripheral component of the world economy. In 1998 the collective GNP of the fourteen states comprising SADC, with a combined population of 196m., stood at just $186bn. (Figure 15.1; Table 15.1). That this is substantially less than the GNP of the Netherlands ($300bn.), a country of only 15m. people, underlines the relative insignificance of southern Africa in global terms. In terms of both trade and foreign direct investments, southern Africa's importance to the rest of the world is not only small but also declining. Whilst one country, South Africa, and certain commodities, like gold, diamonds and coal, are more fully integrated into the world economy, southern Africa is generally perceived to be largely detached from the world trading system. This rather depressing analysis of southern Africa's

273

Figure 15.1 Member States of SADC and SACU

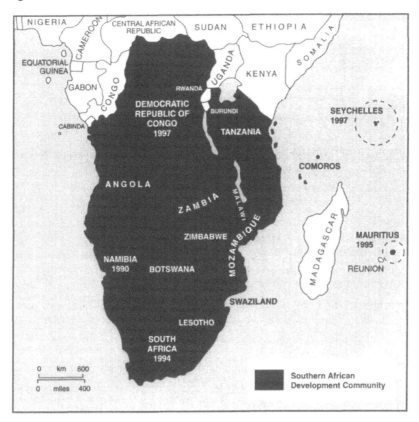

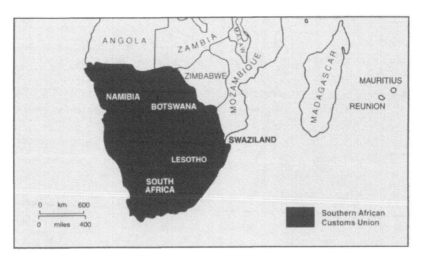

role in the world economy is compounded by its prospects for recovery in the short to medium-term. With a few exceptions, southern Africa's comparative advantage rests in raw materials and cheap, low-paid, unskilled and (particularly outside South Africa) often non-unionised labour. Both of these resources look less attractive in a world economy that is increasingly knowledge-based and driven by technological developments.

Table 15.1   SADC Countries: Basic Economic Indicators

	Land area km^2	Population Millions 2000	GNP US$/ Billions 1998	GNP per capita US$	Human Development	Real GDP growth rates %			
						1980-90	1996	1997	1998
Angola	1,247,000	12.87	4.6	380	160	2.1	12.1	5.9	1.7
Botswana	585,000	1.62	4.8	3070	112	16.9	7.0	6.9	8.3
Congo, D.R.	2,345,409	47.70	5.4	110	141	0.7	0.9	-6.4	-3.5
Lesotho	30,555	2.15	1.2	570	127	3.9	12.7	3.5	-5.8
Malawi	118,484	10.92	2.2	210	159	2.2	12.0	5.3	6.2
Mauritius	1,865	1.16	4.3	3730	59	4.9	6.0	5.2	5.6
Mozambique	790,380	19.68	3.5	210	169	0.1	6.4	6.0	11.6
Namibia	824,269	1.73	3.2	1940	115	-0.6	2.1	2.4	2.6
Seychelles	455	0.91	0.5	7142	66	3.6	1.5	7.9	3.0
South Africa	1,223,201	43.38	136.9	3310	101	1.5	3.1	1.7	0.1
Swaziland	17,000	1.00	1.1	1210	113	6.6	3.9	3.8	2.5
Tanzania	945,000	31.51	7.2	220	156	3.3	4.2	3.3	4.0
Zambia	752,614	10.04	3.2	330	151	1.0	6.5	3.5	-1.8
Zimbabwe	390,757	11.67	7.2	620	130	4.2	7.0	2.0	1.6
SADC	9,271,789	196.34	185.3	943	nd	nd	4.1	2.2	1.7
South Africa as % of SADC	13.2	22.0	73.9	na	na	na	na	na	na

The human consequences of this marginality are profound. About 76m. people, approximately 40 per cent of SADC's entire population, live in extreme poverty with high levels of malnutrition, illiteracy, unemployment, declining life expectancy and limited access to basic social services. According to SADC (2000a), the region has the highest proportion of people subsisting on less than $1 a day in the world. Even in a relatively prosperous country like Botswana, which has experienced sustained economic growth of over 5 per cent per annum throughout the 1990s (chapter 10), 47 per cent of the population live below the poverty line. In South Africa, despite having one of the highest GDP per capita rates in SADC, 44 per cent of its population live in poverty. The figures for

Mozambique and Zambia are significantly worse, standing at 69 per cent and 70 per cent respectively. Furthermore, the level of real economic growth often lags behind the level of population growth (Table 15.1). The United Nations Economic Commission for Africa (2000) estimates that the region requires a GDP growth rate of over 6 per cent to reduce poverty by half by the year 2015. However in the period 1996-2000, SADC GDP growth rates averaged just 2.7 per cent (Ramsamy, 2001). Unless this growth rate improves significantly, poverty levels are likely to remain at best static. According to SADC (2000a), real income per capita today is lower than it was in 1970 for most countries in SADC.

Given the small size of the region's economy, its peripherality in the world trading system and pressures emanating from the world economy, particularly liberalisation as manifested through structural adjustment programmes and international trading agreements like the Uruguay Round and Cotonou, regional economic integration is seen as a means of survival in a rapidly globalising era, as the following comments illustrate:

> Regional integration will prepare us for global competition. If we do not integrate now and put our industries on a more competitive basis, they will be wiped out ... (Joaquim Chissano, President of Mozambique, 2000).

> With the deepening and acceleration of its integration agenda ... the region is increasingly enhancing its capacity and position to minimise the risks of globalisation and take advantage of the opportunities it presents in order to deal with the pervasive problems of extreme and overall poverty (Prega Ramsamy, Executive President of SADC, 2000).

However, the limited size and weak structure of most of the individual economies in SADC can also be used to undermine arguments in favour of promoting regional integration. Regionalism in southern Africa, or anywhere else for that matter, can only be as strong as its constituent parts. Put simply, regional integration is limited by the strength of participating member states and in southern Africa those states are, on the whole, extremely weak, both economically and politically. Two SADC states, Angola and the Democratic Republic of Congo (DRC), can be classified as 'dysfunctional', the latter being close to collapse. Other states, such as Zimbabwe, are in economic and political crisis (chapter 13). This situation has led a number of academics to question the validity of regional integration as a means of promoting development. Christopher Clapham (2001, p. 60) argues that:

... in economic terms, a regional community comprising small and poor African states had little to offer ... regional partners had economies very similar to their own, and the lumping together of a number of (in global terms) tiny African economies did not produce either a viable market for a process of internal industrialisation, or enable them to form any influential bargaining capacity ... with the outside world.

The reservations expressed by Clapham are supported by Herbst who asked the rather blunt question: 'why go through the effort to try to have a relatively small number of extraordinarily poor people trade with each other when the world economy is larger, more populous and growing faster?' (Herbst, 1998, p. 31). The views expressed by Herbst and Clapham highlight a fundamental division amongst those who promote the cause of regionalism in southern Africa. On the one hand, there are those who advocate a regionalism that incorporates interventionist policies designed to protect industries and reduce spatial and structural inequalities (Tsie, 1996; Weeks, 1996; Keet, 1999). On the other hand, there is a group who promote integration so long as it is 'open regionalism' based on competition, comparative advantage and, most importantly, avoids the inward-looking and import-substitution strategies of the 1960s (see, for example, African Development Bank, 1993; Clapham, 2001).

The debate over whether regionalism should be interventionist or market orientated is of particular importance in southern Africa as a result of the unusual degree of inequality in levels of development between South Africa and the other states of the region. The single most important factor determining the nature of southern African regionalism is South Africa's absolute dominance of the region and how to manage that dominance (Gibb, 2001a). The spatial configuration of development established by a particular blend of colonialism, capitalism and apartheid destabilisation is characterised by intense inequality. In 1998, 73 per cent of the total GDP of southern Africa was produced in South Africa. Not surprisingly, therefore, South Africa dominates completely intra-regional trade (South Africa Update, 1999). As a result of the Republic's relatively sophisticated manufacturing sector, it is in a very good position to export a wide variety of consumer and producer goods to a region that, with the exception of Zimbabwe, lacks a developed industrial economy. Southern Africa's striking inequalities and South Africa's hegemonic position within the sub-continent highlight the need for regional integration to incorporate some form of redistributive and compensatory mechanisms that favour the least developed. Clearly, the potential for conflict between those forces advocating affirmative action to assist underdeveloped countries and those advocating open regionalism based on comparative advantage is considerable. Set against this background of inequality and dominance, the remaining section of this chapter examines the region's most important integrative institutions: SADC and the Southern African Customs Union (SACU).

The Southern African Development Community (SADC)

SADC is the successor organisation to the Southern African Development Co-ordination Conference (SADCC) whose primary objective was to reduce the region's dependence on apartheid South Africa. Established by the Lusaka Declaration in April 1980, nine countries - Angola, Botswana, Lesotho, Malawi, Mozambique, Swaziland, Tanzania, Zambia and Zimbabwe - declared their commitment to 'pursue policies aimed at economic liberation and the integrated development of national economies' (Amin et al., 1987). SADCC's geostrategic objective was to divide southern Africa by isolating apartheid South Africa. Initially SADCC established a reputation for success, particularly in attracting large amounts of development aid. By 1992 over $4bn. had been raised, 82 per cent of which was spent on the transport and communications sector (Simon and Johnston, 1999). SADCC was also successful in promoting and preserving regional solidarity in the face of considerable ideological diversity amongst member states and South Africa's attempts to destabilise the organisation and its members. However since 1990 critics have not been reluctant to draw attention to some of the failures of SADCC, particularly over its inability to reduce dependence on South Africa and promote equitable regional growth. During the apartheid era SADCC relied upon South Africa to supply a considerable proportion of its imports. By the start of the 1990s, South Africa remained an important trading partner for all SADCC states except Angola and Tanzania. According to the Development Bank of Southern Africa (1990), South African imports dominated the markets of Lesotho and Swaziland (both 90 per cent) and Namibia (75 per cent), and were the single most important source for Malawi, Zambia and Zimbabwe. That SADCC was not able to reduce its dependence on South Africa is not altogether surprising given the relative size, diversity, sophistication and regional competitiveness of the South African economy. In addition, Pretoria's policy of regional destabilisation enforced SADCC's dependence on South Africa (Gibb, 1994).

Throughout the 1980s South Africa, in response to what it perceived to be a communist-inspired total onslaught against white minority rule, adopted the 'total strategy'. This aimed to defend apartheid by destabilising those southern African countries hostile to the apartheid regime. Although the total strategy incorporated all elements of state policy, including cross-border military raids and the support of resistance/terrorist movements such as Renamo in Mozambique and Unita in Angola, one of its most effective policy instruments involved the manipulation of the region's transport infrastructure. By undermining the transport infrastructure of SADCC, particularly the region's ports and railways, Pretoria successfully and artificially deepened southern Africa's dependence on South Africa (Hanlon, 1986).

SADCC also failed in its endeavours to promote meaningful progress towards regional economic integration. SADCC's integrative strategy deliberately avoided competition, duplication and free market strategies in favour of project co-ordination, with a particular emphasis on large-scale transport infrastructure projects. SADCC's priorities were to manage, plan and co-ordinate regional integration and, in so doing, it totally rejected the neo-classical open regionalism model. The limitations of this 'project co-ordination' approach were recognised in the 1992 SADCC Theme Document (SADCC, 1992a, p. 22):

> The level of integration which currently exists among the ten SADCC member states ... remains extremely modest ... regional project co-ordination ... has been recognised as having only a limited impact in promoting deeper or wider co-operation and integration.

The principal manifestation of this failure to integrate was the persistently low level of intra-regional trade, standing at 5 per cent or less in the period 1980-94. This lack of intra-regional trade was the result of several contributory factors, of which the lack of complementarity between national economies was prominent: most countries export agricultural and mineral commodities for which the major markets are in developed industrial economies. Other constraints to intra-regional trade include uncompetitive industries, small markets, inadequate infrastructure, under-developed financial and capital markets, and a lack of modern technology and skills (Ramsamy, 2000).

SADCC's anti-apartheid stance and its resistance to South African regional hegemony became the organisation's principal *raison d'être* and source of unity in the period 1980-90. Several factors contributed towards SADCC's evolution to a Development Community (SADC), including the ending of the Cold War, enhanced European integration and a fear of a fortress Europe and, most importantly, the constitutional negotiations within South Africa to end apartheid. In August 1992, the Development Community replaced the Co-ordination Conference. The SADC Treaty (1992b) calls for the development of policies aimed at the progressive elimination of obstacles to the free movement of goods, capital, labour and services. In short, its integrative strategy incorporates the basic elements of the customs union approach, the foundation of which is a free trade area (Gibb, 2001a). At the same time, however, SADC continues to recognise the threats presented by regional inequalities arising from an uneven distribution of the benefits attributable to integration, noting that 'trade liberalisation (is) to be complemented by compensatory and corrective measures, orientated particularly towards the least developed countries' (SADC, 1992b, p. 22).

SADC is a dynamic organisation that has experienced several enlargements, including Namibia (1990), South Africa (1994), Mauritius (1995), the DRC and the Seychelles (both 1997). Although the fourteen states comprising SADC are geographically contiguous, the addition of new members has made the organisation significantly more diverse. In terms of political stability, economic infrastructure and resources, language and colonial legacies, SADC exhibits considerable internal diversity. This diversity is a particular challenge to an organisation like SADC whose political institutions and decision-making processes are widely regarded as inadequate (Sidaway and Gibb, 1998).

The organisational structure of SADC is complex but can be simplified into two basic components. First, there are those institutions, the 'Summit' and 'Council of Ministers', responsible for overall policy direction. The Summit is made up of the Heads of State or Government and the Council is comprised of ministers from each of the fourteen member states, usually those concerned with economic/trade issues. The Council advises the Summit on policy matters and approves SADC policies, strategy and work programmes. However both the Summit and Council meet infrequently, perhaps once or twice a year under normal circumstances. Decisions in both the Summit and Council are taken by consensus, with each state having a *de jure* right of veto. The Summit is characterised by 'an atmosphere of ritualised diplomatic politeness and protocol' (Sidaway and Gibb, 1998, p.168) where diplomatic etiquette prevents substantial critique of other member states. The real horse-trading takes place in the Council, again on the basis of consensus.

As is common in most regional organisations, there is an almost continuous conflict between national and regional interests which, in the case of SADC, is exacerbated by having a very limited secretariat of genuinely regional civil servants. In 2000 the SADC Secretariat comprised only 35-40 staff, plus technical and secretarial support staff. The level of supranational governance in SADC is therefore limited, with all key decisions requiring consensus and unanimity.

The Treaty establishing SADC (SADC, 1992b), signed in Windhoek in August 1992, sets out an ambitious integrative programme, the most important aims of which are to:

- harmonise the political and socio-economic policies and plans of member states;
- develop policies aimed at the progressive elimination of obstacles to the free movement of capital and labour, goods and services;
- promote the co-ordination and harmonisation of the international relations of member states.

However the Treaty does not bind member states to specific actions, outcomes or targets but sets out a list of general objectives. These objectives are met through the use of protocols attached to the Treaty. Each protocol, for example the Trade Protocol or the Shared Water Courses Protocol, has to be approved by the Summit, negotiated by consensus in the Council and then ratified by individual member states. Clearly, national interests have priority over regional interests. Commenting on the weakness of SADC's institutional infrastructure, Ramsamy (2000, p. 5) noted:

> The delay in realigning institutions inherited from the Coordination Conference to make them consistent with the new mandate of regional integration ... has thwarted desires to build the required institutional capacities ... for effective and efficient management of the integration process.

Notwithstanding these institutional constraints, SADC has passed protocols on immunities and privileges for SADC personnel, drug trafficking, a southern African energy pool, mining, education and training, the integration of transport communications and, most significantly, a trade protocol. However, many of these protocols await ratification by member states.

The SADC Free Trade Area (FTA)

In August 1996 eleven out of the then twelve SADC members (all but Angola) signed a 'Protocol on Trade' which committed signatories to establish a FTA (SADC, 1996). For the Protocol to enter into force it had to be ratified by at least a two-thirds majority, which was achieved on 25th January, 2000. The initial participants are South Africa, Botswana, Lesotho, Malawi, Mauritius, Mozambique, Namibia, Swaziland, Tanzania, Zambia and Zimbabwe. War-torn Angola and the DRC decided not to join the FTA, as did the geographically remote islands of the Seychelles. The two most important objectives of the SADC Trade Protocol (SADC, 2000b) are:

- the further liberalisation of intra-regional trade in goods and services on the basis of fair, mutually equitable and beneficial trade agreements;
- the establishment of a free trade area in the SADC region.

The basic characteristics of the SADC FTA are:

- lower tariffs (immediately) on all 'non-sensitive' trade to under 18 per cent;
- the liberalisation of 85 per cent of all intra-regional trade by 2008;
- the liberalisation of the remaining 15 per cent of 'sensitive' products by 2012;
- the promotion of asymmetry whereby South Africa (and therefore states belonging to the Southern African Customs Union) will provide wider market access over a shorter time-period, resulting in all duties at 25 per cent and above being eliminated in five years;
- an aim to increase intra-regional trade from 25 per cent (see below) to 35 per cent and to facilitate foreign trade with the region as a whole.

The extent to which the FTA will increase SADC's trading profile and generate welfare benefits will depend, in part, on the relative strengths of trade diversion and trade creation. It has long been recognised that a free trade area does not necessarily represent a step towards more free trade or enhanced welfare. If SADC imports of efficiently produced products from countries outside the region are displaced by more expensive imports from South Africa, then increased intra-regional trade could lower total trade and impose welfare losses as a result of this diversion. Conversely, trade creation may occur as a result of high-cost domestic production being replaced by cheaper imports sourced from within the FTA. However, both trade creation and diversion have the potential further to increase regional inequalities.

The FTA will have a number of important short- and long-term impacts. In the short-term, the FTA will result in a loss of customs revenues, upon which many of the poorer SADC countries depend. Malawi and Mozambique, because most of their imports come from other SADC countries, could lose up to 90 per cent of customs revenues, and Zimbabwe and Zambia could lose 41 per cent and 25 per cent respectively (Gibb, 2001b). The FTA could therefore result in regional governments having to raise taxes by as much as 10 per cent in order to offset the loss of customs revenue. In addition certain sectors, for example agriculture in Mozambique, glass in Zimbabwe and leather products in Malawi, are likely to suffer considerable losses as a result of free trade (United States Department of State, 2000).

More important is the long-term impact of a FTA on the existing and unusually high levels of regional inequality amongst SADC member states. Many commentators expect the trade imbalance between South Africa and the other member states of SADC to grow further as a consequence of the FTA. Theoretical and empirical evidence suggests that under conditions of free trade the larger and more successful states benefit at the cost of weaker states, especially in a regional organisation comprised primarily of developing countries (Asante, 1997; Gibb, 2001a).

Since South Africa's accession to SADC in 1994, trading patterns provide evidence that appears to support the thesis that freer trade within southern Africa will enhance the region's inequalities. South Africa's membership of SADC has had a profound impact on the organisation in general and intra-regional trade in particular, raising the level of internal trade from below 5 per cent in the early 1990s, to 17 per cent in 1995 and 25 per cent in 2000. The initial boost given to intra-regional trade following South Africa's membership was more statistically apparent than real. South Africa already had extensive trading relations throughout southern Africa before joining SADC that, on membership, were re-classified as intra-regional. None the less, intra-regional trade has experienced real and considerable growth, expanding 28 per cent in the five years up to 2000. As observed by Mayer and Thomas (1997, p. 26):

> In contrast to the poor trade potential amongst many southern African countries, there is a great deal of complementarity between the South African economy and its neighbours. In some ways the SADC countries' trade with the Northern countries is similar to their trade with South Africa. The African market is clearly very important for South Africa.

Although intra-SADC trade has grown significantly since 1994, it has not grown in a balanced fashion. Indeed, the most striking feature of South Africa-SADC trade is the high ratio of exports to imports, which stood at 4.2:1 in 1994 and has widened to 7.2:1 in 1998-99. Thus, while 86 per cent of intra-regional imports are supplied by South Africa, South Africa's trade with SADC states amounts to just 7 per cent of exports and 2 per cent of imports (Mills and Sidiropoulos, 2001).

The SADC FTA is attempting to address the issue of some states doing better than others by adopting an 'asymmetrical approach' to trade liberalisation, whereby South Africa (and therefore the Southern African Customs Union) provides better market access over a shorter time-period than other SADC signatories to the FTA. The redistributive mechanism adopted by SADC to tackle the issue of inequality is, therefore, market-led and based on the assumption that non-SACU industries will be able to exploit the benefits offered by a short-term preferential tariff regime. However, it is unlikely that this asymmetrical approach to trade liberalisation will be enough to prevent the South Africa-SADC trade imbalance growing wider as a result of the FTA. In the long-term, the asymmetrical nature of the South Africa-SADC trading relationship may prove unsustainable.

The Southern African Customs Union

The Southern African Customs Union Agreement (SACUA) between Botswana, Lesotho, Namibia and Swaziland (BLNS) and South Africa is widely regarded as the most effectively functioning trade agreement in Africa (Gibb, 1997). It provides for the duty-free movement of goods and services between member states, a common external tariff (CET) against the rest of the world and a revenue-sharing formula. SACU is therefore a sophisticated form of regionalism representing a level of integration significantly deeper than a free trade area.

The origins of SACU date back to the 1889 Customs Union Convention between the British Colony of the Cape of Good Hope and the Boer Republic of the Orange Free State. In 1910, following the Union of South Africa, a Customs Union Agreement was signed between South Africa, Bechuanaland (Botswana), Basutoland (Lesotho) and Swaziland. Namibia has been a *de facto* member of the Union since 1915, when South Africa took over the administration of German South West Africa.

The original 1910 Agreement established a revenue-sharing formula based on the levels of external trade in the period 1907-10. Thus, South Africa received 98.7 per cent of joint revenue whilst the High Commission Territories (HCTs) received collectively 1.3 per cent; Bechuanaland 0.27 per cent, Basutoland 0.88 per cent and Swaziland 0.15 per cent. Both the United Kingdom and South Africa, the principal negotiators of the 1910 Agreement, envisaged that the HCTs would eventually be incorporated into the Union of South Africa. As a result, the ability of the HCTs to manage their economies was not really considered, as economic development beyond resource extraction for overseas markets was not envisaged. However the 1910 Agreement came under considerable strain following South Africa's decision in 1925 to adopt a policy of import substitution which resulted in highly protective tariff barriers being imposed around SACU in order to stimulate economic development in South Africa. For the HCTs, protective tariff barriers promoted trade diversion and welfare losses as consumers were forced to purchase high-cost South African produce. In addition, industrial development resulting from the policy of import substitution was polarised on South Africa. Thus, from the 1930s onwards, pressure to renegotiate the 1910 Agreement intensified. Negotiations for a revised agreement commenced in 1963 but were delayed pending the independence of the HCTs: Botswana and Lesotho in 1966 and Swaziland in 1968.

SACU therefore has its origins firmly embedded in a colonial policy that aimed to incorporate the HCTs into South Africa. Furthermore, until the late 1960s, the HCTs were under colonial rule and had little influence in the organisation or policies of SACU. In 1969 SACU was renegotiated. Unlike the 1910 Agreement which, to all intents and purposes, ignored economic development and integration, the 1969 Agreement set out clearly the Union's role to promote economic development throughout all member states on the basis of equity:

> to ensure the continued economic development of the customs union area as a whole, and to ensure in particular that these arrangements encourage the development of the less advanced members of the customs union and the diversification of their economies, and afford all parties equitable benefits arising from trade among themselves and with other countries (Republic of South Africa, 1969).

In order to achieve this ambitious goal the SACU Agreement set out a number of provisions, the most important of which was a new revenue-sharing formula to apportion the monies collected by the Common Revenue Pool (CRP). Revenue from the Pool was shared out according to a complex formula:

- the share of the total CRP paid to BLNS depended on the level of their own imports, customs duties, excise, import surcharges and sales duties as a percentage of the totals of these amounts for the Customs Union as a whole;
- this sum was then augmented by a 'compensation rate' of 42 per cent in order to offset the disadvantages of small countries belonging to a customs union dominated by a more developed industrial economy (see below);
- in 1976 the formula was amended to provide a 'stabilisation factor' that ensured the BLNS received at least 17 per cent of the total value of all their imports and other duties paid. This sum therefore bore no relation to the absolute size of the CRP.

The stabilisation factor was the binding element within the formula and increased the effective rate of compensation paid to BLNS to 77 per cent (Gibb, 1997). The amount retained by South Africa was the residual after payments had been made. Negotiations to restructure the 1969 Agreement commenced in November 1994 in Windhoek and an agreement in principle was reached in September 2000. The BLNS were dissatisfied with the 1969 Agreement for several reasons. First, they argued that the revenue-sharing formula did not compensate them adequately for the costs of having to purchase uncompetitive high-cost South African goods protected from world competition by a high CET. Second, the benefits of this policy, in terms of enhanced industrialisation, were felt primarily in South Africa. SACU promoted the polarisation of industrial activity in South Africa, a trend accentuated in the 1980s when Pretoria promoted a generous incentive scheme designed to attract investments to its black 'homelands'. Third, the 1969 Agreement obliged the BLNS countries to apply a tariff regime unilaterally determined by South Africa. Although BLNS had to be consulted, South Africa could impose, amend or terminate tariffs unilaterally.

The BLNS states did, however, derive substantial and significant benefits from the 1969 SACUA. As Maasdorp and Whiteside (1993) point out, they could have withdrawn from SACU at any time and in 1990, after gaining independence, Namibia chose to join. The most important benefit for the BLNS states was the very significant contribution to state revenue provided by the CRP. In 1995/6, SACU receipts accounted for between 16.3 per cent and 50.6 per cent of BLNS government revenue (Table 15.2).

Table 15.2   SACU Receipts as a Percentage of Total Government Revenue

	1998/9	1990/1	1991/2	1992/3	1993/4	1994/5	1995/6
Botswana	11.5	13.2	19.0	21.9	15.9	16.2	16.3
Namibia	31.9	22.3	26.3	25.3	25.4	26.4	30.1
Swaziland	37.2	44.7	43.7	36.9	46.3	47.2	50.1
Lesotho	41.5	43.5	43.8	56.5	53.3	53.2	50.6

South Africa's principal dissatisfaction with the 1969 Agreement was the increasing costs of supporting the stabilisation factor. The BLNS states' share of the CRP had grown from 2.6 per cent in 1969-70 to 32 per cent in 1991/2. Most importantly, however, South Africa decided in 1992-94 to pursue a policy of tariff liberalisation. Under the GATT Uruguay Round and subsequent trade agreements, most notably the European Union-South Africa Trade, Development and Co-operation Agreement (chapter 16) and the SADC FTA, the size of the CRP was set to fall dramatically. Over 40 per cent of SACU's imports originate from within the EU and will, following a twelve-year transition period, enter the Customs Union without duty. Under the existing agreement, however, the impact of these reductions would not have been passed on to the BLNS states which were protected by the stabilisation factor.

Whilst the above figures demonstrate clearly the crucial importance of the 1969 SACUA to the BLNS countries, the Agreement was also vital to South Africa in terms of its manufacturing export market and overall trade balance. The biggest benefit to South Africa arose from securing BLNS markets for its internationally uncompetitive goods (i.e. trade diversion). SACU accounts for approximately 25 per cent of all South Africa's manufacturing exports. As Table 15.3 illustrates, SACU represents a significant proportion of South Africa's African export and import trade, standing at 61.5 per cent and 58.5 per cent respectively.

Table 15.3   South Africa's Trade with SACU and SADC Countries, 1995

	% of exports to Africa	% of imports from Africa
Botswana	18.6	9.9
Namibia	18.1	24.8
Swaziland	12.6	20.1
Lesotho	12.2	3.7
SACU subtotal	61.5	58.5
Non-SACU SADC		
Zimbabwe	11.0	17.9
Mozambique	6.3	1.6
Zambia	5.2	3.4
Malawi	2.8	3.2
Mauritius	2.4	0.3
Dem. Rep. of Congo	1.6	6.4
Angola	1.4	0.3
Tanzania	0.8	0.3
Seychelles	0.4	0.1
Non-SACU SADC	31.9	33.5
SADC	93.4	92.0

Clearly SACU is of critical importance to all SACU countries, including South Africa. SACU's longevity reflects the benefits both parties received from the Agreement. However, the BLNS perception was that the 1969 revenue sharing-formula did not adequately compensate them for the detrimental affects of SACU membership, particularly trade diversion and industrial polarisation. South Africa, on the other hand, perceived the 1969 Agreement to be unsustainable and too generous to the BLNS states.

Renegotiating the 1969 SACUA, originally timetabled to take just two years, proved to be far more acrimonious and complex than originally expected. The negotiations were complicated by the impact and consequences of other trade agreements, most notably the free trade areas with Europe and other SADC states, the Lomé and Cotonou negotiations between the BLNS and the EU, and the implementation of commitments made under the Uruguay Round. Negotiations focused on three issues. First, the need to change the revenue-sharing formula used to allocate the CRP. Second, a BLNS desire to institutionalise SACU by creating a secretariat with representatives from all member states. Finally, the BLNS states demanded a more democratic arrangement that allowed them a say in the decision-making process affecting the SACU tariff regime.

In September 2000 the negotiating teams agreed an outline framework for a new SACU. The most sensitive and contentious issue concerned the revenue pool and a new revenue-sharing formula based on two components. First, income from customs tariffs collected via the CET will be distributed according to intra-SACU trade levels. Thus, the more one country imports from the others the higher will be its share. On current trade patterns this formula will favour the BLNS countries. However, as tariffs fall the amount collected via customs is set to reduce dramatically. The second part of the revenue-sharing formula concerns sharing income from excise duties. Approximately 15 per cent of excise duty income will be set aside for development purposes, shared on the basis of an inverse per capita income relationship, with low per capita income countries getting the largest share. The remaining 85 per cent of income derived from excise duties will be distributed according to GNP. Structured in this fashion, the new revenue-sharing formula retains a significant element of subsidy that favours the BLNS.

It has also been agreed to establish a new institutional structure comprising a Ministers' Council for decision-making and a new multilateral secretariat or Commission of senior officials. The latter, funded from the revenue pool, will deal with administrative issues such as the setting of tariffs, trade disputes and the management of the revenue-sharing formula. Decisions in both the Ministers Council and Commission will be via consensus.

Conclusion

The principal obstacles facing regional economic integration in southern Africa arise from the significant inequalities in the levels of development between South Africa and the other states of the region. The most striking feature of intra-regional trade in the period 1994-2000, during which time the sub-continent experienced substantial economic liberalisation, was the very high ratio of South African exports to imports, resulting in Pretoria holding a massive trading surplus with the region. Under conditions of unrestricted free trade the SADC FTA could exacerbate this trend still further and damage the economies of many SADC states. As Mayer and Thomas (1997, p. 28) observe, 'it is widely held that such a surplus is unsustainable, both within South Africa and in the rest of the region ... impoverished countries do not make for strong trading partners'.

Several previous attempts to promote regional economic integration in Africa, such as the East African Community, have faltered on the issue, real or perceived, that some countries benefit at the expense of others. SACU, by supporting a compensatory payments mechanism that favours the smaller BLNS states, addresses the issue of inequality in an interventionist and corrective manner. The 1969 Agreement established a compensation rate and stabilisation factor whilst the new SACUA has a development fund.

SADC's strategy to redress the current inequitable trading relationship is market-driven and centred on eliminating all obstacles to the free movement of intra-regional trade, albeit with an asymmetrical timetable that favours non-SACU SADC countries. Given the evidence that free market conditions tend to favour stronger economies at the expense of the least developed, the SADC FTA would appear to be an inadequate instrument to address the prevailing obstacles to promoting balanced intra-regional trade. Inevitably, demands that trade should be 'fair' as well as 'free' will lead to pressures for more co-ordinated and interventionist policies designed to promote growth in the least developed countries. However, South Africa's ability to promote a significant transfer of resources beyond the BLNS states is extremely limited. In an era dominated by multilateral liberalisation and tariff reductions, it looks likely that economic integration in southern Africa will remain handicapped by the issue of how to address the problems associated with the region's profound inequalities.

References

African Development Bank (1993), *Economic Integration in Southern Africa*, Volumes I, II, III, Biddles, England.

Amin, S., Chitala, D. and Mandaza, I. (1987), *SADCC: Prospects for Disengagement and Development in Southern Africa*, Zed Books, London.

Asante, S. (1997), *Regionalism and Africa's Development*, Macmillan Press, Basingstoke.

Chissano, J. (2000), Official SADC Trade, Industry and Investment Review, *SADC Review*, SADC, Gaborone.

Clapham, C. (2001), 'The Changing World of Regional Integration in Africa', in C. Clapham et al., op. cit., pp. 59-71.

Clapham, C., Mills, G., Morner, A. and Sidiropoulos, E. (eds) (2001), *Regional Integration in Southern Africa*, South African Institute of International Affairs, Johannesburg.

Development Bank of Southern Africa (1990), 'Financial Resources and Capital Investments in the Common Monetary Area', paper presented to a conference on 'New post-apartheid South Africa and its neighbours', Maseru, 9-12 July 1990.

Gibb, R. (1994), 'Regional Economic Integration in Post-apartheid Southern Africa', in R. Gibb and W. Michalak (eds), *Continental Trading Blocs: The Growth of Regionalism in the World Economy*, John Wiley, London, pp. 209-31.

Gibb, R. (1997), 'Regional Integration in Post-apartheid Southern Africa: the case of Renegotiating the Southern African Customs Union', *Journal of Southern African Studies*, vol. 23, pp. 67-86.

Gibb, R. (2001a), 'The State of Regional Integration: The Intra- and Inter-regional Dimensions', in C. Clapham et al., op. cit., pp. 71-87.

Gibb, R. (2001b), *Southern Africa: Integration Quest*, Oxford Analytica Daily Brief, June 6.

Hanlon, J. (1986), *Beggar Your Neighbours*, James Currey, London.

Harvey, C. (1997), *The Role of Africa in the Global Economy: the Contribution of Regional Co-operation, with Particular Reference to Southern Africa*, BIDPA working paper no.11, Gaborone.

Herbst, J. (1998), 'Developing Nations, Regional Integration and Globalism', in A. Handley and G. Mills (eds), *South and Southern Africa: Regional Integration and Emerging Markets*, South African Institute of International Affairs, Johannesburg, pp. 29-40.

Keet, D. (1999), 'Regional Integration and Development in Southern Africa', paper presented to a conference 'Farewell to Lomé', 23-25 April 1999, Konigswinter, Bonn.

Maasdorp, G., and Whiteside, A. (1993), *Rethinking Economic Co-operation in Southern Africa: Trade and Investment*, Occasional Paper, Konrad-Adenauer-Stiffung, Johannesburg.

Mayer, M., and Thomas, R. (1997), 'Trade Integration in the Southern Africa Development Community', in Kritzinger van Niekerk (ed), *Towards Strengthening Multisectoral Linkages in SADC*, Development Bank of Southern Africa, Development Paper no. 33, Halfway House, Midrand.

Mills, G., and Sidiropoulos, E. (2001), 'Trends, Problems and Projections in Southern Africa Integration' in C. Clapham et al., op. cit., pp. 9-13.

Ramsamy, P. (2000), *Poverty Reduction: A Top Priority in SADC's Integration Agenda*, SADC Review, SADC, Gaborone.

Ramsamy, P. (2001), 'SADC: The Way Forward', in C. Clapham et al., op. cit., pp. 33-43.

Republic of South Africa (1969), *Customs Union Agreement and Memorandum of Understanding between the Governments of the Republic of South Africa, Botswana, Lesotho and Swaziland*, Republic of South Africa Treaty Series No. 8/1969, Government Printer, Pretoria.

SADC (1992a), *Towards the Southern African Development Community*, SADC, Gaborone.

SADC (1992b), *Treaty of the Southern African Development Community*, SADC, Gaborone.

SADC (1996), *Protocol on Trade in the Southern African Development Community Region*, SADC, Gaborone.

SADC (2000a), *Regional Human Development Report*, SADC, Gaborone.

SADC (2000b), *The SADC Free Trade Area*, SADC, Gaborone.

Sidaway, J., and Gibb, R. (1998), 'SADC COMESA SACU: Contradictory Formats for Regional Integration in Southern Africa', in D. Simon (ed) *Reconfiguring the Region: South Africa in Southern Africa*, James Currey, London, pp. 164-84.

Simon, D., and Johnston, A. (1999), *The Southern African Development Community*, Briefing Paper New Series No. 8, The Royal Institute of International Affairs, London.

South Africa Update (1999), *Trade between SACU and SADC in the 1990s*, Trade and Industrial Policy Secretariat (TIPS), Johannesburg, South Africa.

Tsie, B. (1996), 'States and Markets in the Southern African Development Community: Beyond the Neo-liberal Paradigm', *Journal of Common Market Studies*, vol. 22, pp. 75-100.

United Nations Economic Commission for Africa (2000), *Annual Report*, United Nations, New York.

United States Department of State (2000), *SADC Trade Protocol*, US and Foreign Commercial Service and US Department of State (www.tradeport.org/ts/countries/safrica/mmr/mark0202.html).

Weeks, J. (1996), 'Regional Co-operation and Southern African Development', *Journal of Common Market Studies*, vol. 22, pp. 100-117.

16 Trade, Aid and Foreign Investment: South Africa, Southern Africa and the European Union

ANTHONY LEMON AND RICHARD GIBB

Introduction: South African Trade Policies in the Apartheid Era

In the late 1920s and early 1930s South Africa embarked on an import substitution growth path. This strategy comprised the erection of import quotas and tariff barriers, and the establishment of parastatal corporations such as ESCOM (electricity) and ISCOR (iron and steel). It was designed to diversify the economy in terms of both production and origin of imports, so reducing South Africa's dependence on Britain. Whilst manufacturing production diversified, ownership became increasingly concentrated as mining conglomerates became involved in manufacturing. Imports of capital and intermediate goods increased, while locally produced consumer goods were only competitive in the domestic market thanks to import protection.

South Africa was a founding member of the General Agreement on Tariffs and Trade (GATT) in 1948. However while other GATT countries liberalised their trade policies through bilateral agreements or creating new trade blocs, domestic political considerations arising from apartheid policies led to the maintenance of protectionist policies. By the 1970s growth had begun to slow, in part because of worldwide recession following the OPEC action to increase oil prices, and the negative consequences of the import substitution programme were becoming increasingly apparent. These included a significant anti-export bias and considerable import dependence, especially on capital goods. The report of the Reynders Commission (South Africa, 1972) led to the introduction of substantial export incentives, but these did little to increase exports, perhaps because they were insufficient to outweigh the cost to exporters of an over-valued exchange rate driven by strong gold exports. It was only in 1983 that serious attempts to dismantle import controls began.

When Chase Manhattan and other banks refused to roll over short-term loans in 1985, South Africa responded by introducing an import surcharge and intensifying previous export promotion policies including devaluation and the reintroduction of a dual exchange rate. In 1990 the General Export Incentive Scheme (GEIS), a cash subsidy paid to exporters on the basis of value added and local content, was introduced. This tended in practice to reward companies already exporting but did little to encourage new firms to export.

Post-apartheid Trade Liberalisation

It was the white-minority government that initiated trade liberalisation in the early 1990s in the context of the GATT Uruguay Round. Negotiations with GATT led to the phasing out of GEIS over five years, but as a result of a re-prioritisation of the national budget and accumulating evidence that the programme was not operating optimally, the process was accelerated and GEIS was finally discontinued in 1997. Other trade-related measures which contravened World Trade Organisation (WTO) rules, such as support for the clothing industry and local content requirements in the motor industry, have also been abolished. More than 12,000 tariff lines are being reduced to less than 6,000, and over 200 tariff rates standardised to just six ranging between 0 and 30 per cent (Hirsch and Hanival, 1998; Roberts, 2000). The average level of tariffs has been reduced by about one-third in five years, and tariffs on many products have been reduced to below WTO rates. Trade liberalisation is intended to increase competition in the domestic market and so dampen inflationary pressures, and to encourage exports by forcing manufacturers to become more competitive and reducing the anti-export bias.

The government's Growth, Employment and Redistribution strategy (GEAR) identifies growth of manufactured exports as an essential ingredient of economic and employment growth. A series of industry analyses has been undertaken to identify their competitive strengths and weaknesses and to design measures to launch key sectors on a globally competitive growth path. A range of WTO-compatible supply-side measures have been put in place to facilitate industrial restructuring, technology upgrading, investment and export promotion, and the development of small, medium and micro-sized enterprises (SMMEs). The Department of Trade and Industry (DTI) has also introduced matching grants for outward-selling trade missions, exhibition assistance and primary export market research.

Substantial devaluation of the rand in 1996, 1998, 2000 and again in 2001 in response to fallout from the Zimbabwe crisis, weaknesses in other emerging markets and offshore profit repatriation and dividend payments by South African companies now listing overseas has improved the competitiveness of South African exports. The dollar/rand exchange rate has slipped from R4.6 in 1997 to R11.4 at the end of 2001. However it increases the cost of imported machinery and high-technology goods and threatens to undermine the government's successful reduction of inflation since 1994.

Trade liberalisation policies reflect orthodox trade theory which argues that liberalisation brings efficiency gains derived from a country producing and exporting according to its comparative advantage. It is assumed to improve allocative efficiency, enhancing gains from trade with further specialisation in production and exports, enabling greater imports and consumption. Dynamic gains are also envisaged: increasing diffusion of knowledge and technology, increasing competition, more production in sectors with higher income and price elasticities and increased investment. However Roberts (2000) questions whether, in the face of pervasive market failures and imperfect competition, industry will respond as predicted.

Export Performance and Trade Liberalisation

South Africa has a relatively open economy which is highly dependent on foreign trade. Despite rapid industrialisation, imports are generally a similar proportion of national income to their average of 24 per cent in the 1930s. In the latter years of apartheid imports were significantly below the level of exports, owing to the need to finance capital repayments overseas, especially in the years 1985-93. Intermediate and capital goods including machinery, chemicals and motor vehicles are the leading imports, and periods of growth tend to lead to sharp rises in imports. Exports have been dominated historically by minerals, and collectively, if steel and other metal products are included, they continue to account for nearly half South Africa's exports, but the share of gold has fallen from 36.4 per cent in 1979 to 15.7 per cent in 1999. Whilst manufactured goods now account for about two-fifths of total exports many of these are lightly processed agricultural and mineral products.

The strong export performance of the motor industry, machinery and electrical machinery appears to present evidence of export growth through liberalisation, but government intervention, in the shape of the Motor Industry Development Programme, has played an important role (van Seventer, 2000), and a new initiative is designed to help smaller automotive components companies find international partners. Catalytic converters have been particularly successful, with South Africa capturing 9 per cent of the world market in 2000 and production expected to double by 2002 (*Business Day*, 2000a). The expansion of armaments export, South Africa's second largest manufacturing export since 1997, has built on the foundations laid in the apartheid era when South Africa faced UN sanctions on arms imports.

South Africa's comparative advantage remains in primary commodities and the first stages of their beneficiation. The success of the minerals-related sub-sector confirms orthodox trade theory, as it represents specialisation in areas of comparative advantage. It is this sector, however, which is responsible for the capital and skill intensive nature of South Africa's exports, which contradicts predictions that South Africa's comparative advantage should be in unskilled labour-intensive products, given its low skills base and high unemployment levels (Roberts, 2000). Also contrary to the export-led growth model is the contraction of many sub-sectors such as transport equipment, beverages and footwear, which have improved their export performance. Trade in these sub-sectors does not appear to have stimulated growth, but only ameliorated decline (ibid.). Government intervention has been significant in sub-sectors in which a major part of production is exported, including iron and steel, non-ferrous metals and chemicals. Trade is also affected by international agreements which link market-sharing to the licensing of technology from large TNCs, as in the chemical industry, and by the global sourcing strategy of TNCs, as in the motor industry (ibid.).

The relatively high cost of unskilled labour in South Africa helps to explain the conjunction of trade liberalisation and rising unemployment (Tsikata, 1999). Other factors are government intervention to reduce the cost of capital to the minerals-related sub-sectors and the complementarity between cheap energy and capital-intensive processes. Fine and Rustomjee (1996) argue that the minerals-energy complex (MEC) has driven industrialisation in South Africa based on linkages between manufacturing and mining with high levels of government intervention (chapter 3). Such experience is consistent with that of the east Asian NICs, where trade performance was based on differential patterns of integration with the world economy as part of a coherent industrialisation strategy, not trade liberalisation.

Directions of Trade and Trade Agreements

South Africa's trade patterns are highly diversified geographically. The share of South African exports going to the USA had declined from 18.7 per cent in 1979 to 8.3 per cent in 1998, and that to the UK from 17.4 per cent to 6.1 per cent, with Japan, Germany and Zimbabwe taking similar shares in 1998. Sources of imports are likewise diverse: Germany leads with 13.4 per cent in 1998, followed by the USA (9.5), the UK (6.6), and Italy (4.3). The European Union (EU) is South Africa's most important export market, followed by the rest of Africa. Exports to Africa tripled between 1992, when sanctions were lifted, and 1999. SADC countries (chapter 15) account for eight of South Africa's ten leading African markets, which are (in rank order for 1998): Zimbabwe, Mozambique, Zambia, Malawi, Angola, Kenya, Mauritius, Tanzania, the DR Congo and Ghana. These countries account for 85 per cent of total African exports. Trade with Nigeria more than doubled in 1998-99, making Nigeria South Africa's major trading partner in West Africa (*Southscan*, 2000).

This diversified trade pattern is reflected in official policies that are attempting to maintain and strengthen trade relations on many fronts. The most important, with the EU, are discussed separately below. The US wants to develop its Africa Growth and Opportunity Act, which is expected to lead to substantial growth in clothing and textile exports, into a full bilateral trade agreement with South Africa (*Business Day*, 2000b). Canada granted South Africa preferential tariff rates in 1994, although textiles, clothing, footwear and some steel goods are excluded. Several countries including Norway, the USA, Japan and Hungary have granted South Africa Generalised System of Preferences (GSP) facilities. South Africa's first trade agreements with India and China were made in 1994 and 1996 respectively. Both are limited in scope, that with China providing for the mutual extension of most-favoured-nation status. South Africa's decision in 1996 to cut diplomatic ties with Taiwan in favour of the People's Republic was clearly influenced by trade considerations, heightened by the impending handover of Hong Kong to China (Lemon, 2000a). A proposed free trade agreement with Brazil has aroused some anxiety in the farming sector, heightened by Brazil's membership of the Mercasur trade bloc which includes Argentina and Uruguay (*Business Day*, 2000c). Further preferential trade agreements with Nigeria and Libya are under consideration (*Business Day*, 2000d).

The Indian Ocean Rim

The Indian Ocean Rim Association for Regional Co-operation (IOR-ARC) was established in 1997 by fourteen states, with possible extension to seven more. At present the IOR-ARC represents an advanced form of co-operation but stops short of market integration. However by 2003 IOR members of the Association of South-East Asian Nations (ASEAN) will have free trade arrangements in the IOR and by 2004 the SADC members of IOR-ARC will be similarly structured, making a free trade scenario within the IOR almost irresistible (Breytenbach, 1999). IOR-ARC is a welcome example of home-grown regional co-operation, assisted by Indian economic liberalisation since 1991 and South African democratisation since 1994. Although extremely diverse and geographically extensive, the IOR is mainly served by the English language, has good ports and banking institutions, some highly competitive economies such as Singapore, strong agricultural sectors in Australia and South Africa, and large markets in India and Indonesia (Moodliar, 1997). It represents about 12 per cent of world trade, of which only one-fifth currently takes place within the region (Breytenbach, 1999).

South Africa has expressed concern that its fellow members of SADC, and African countries generally, should benefit from its participation in IOR-ARC. It may not want to rush into trade integration unless and until the SADC and SA/EU free trade experiences have proved positive. At present it has an overall trade deficit with the IOR-ARC, primarily with oil-exporting countries. Its trade with the region in both directions consists mainly of natural resource products, and is growing much faster than South Africa's trade generally, a trend which leads Holden and Isemonger (1999) to suggest that a re-evaluation of trade priorities may be called for; others are more sceptical (Ahwireng-Obeng, 1997).

South Africa, Developing Countries and South-South Relations

Since it has moved from pariah to participant, South Africa has actively voiced the concerns of developing countries in the WTO. It argues that *developing countries* have the potential for rapid growth as barriers to international trade decrease, but that they cannot be seen as a single category for which a 'one-size-fits-all' recovery model is appropriate and need help based on their specific needs. To this end, South Africa advocates that developing countries should have a stronger voice in multilateral negotiations, where they should seek far-reaching structural adjustment in the economies of *developed* countries. Such restructuring would reduce current protective and support measures including export subsidies which maintain inefficient agriculture and 'grandfather' industries such as clothing and textiles, footwear and leather goods, and thereby allow a relocation of production and investment to developing countries which possess comparative advantages in these areas. South Africa is also concerned at the proliferation of health and safety standards which can easily become obstacles to the exports of developing countries.

Recognising the importance of alliances and coalition-building in multilateral negotiations, South Africa is actively seeking to build common approaches on key issues within the southern African region, and with other African and developing countries, to strengthen the developmental dimension in negotiations. Brazil, Egypt, India and Nigeria have been identified as important strategic players in the South, and South Africa is working to establish a 'G-South' forum to draw these countries together and promote a common agenda in the WTO to unlock the growth potential of the developing world (ITEDD, 2000).

Foreign Direct Investment (FDI) and Portfolio Investment

Scarcity of capital is the most important problem facing the South African economy. With one of the lowest domestic savings rates in the world, South Africa is very dependent on foreign capital flows to finance investment in productive capacity (Lemon, 2000b). It experienced a steady outflow of capital between 1985 and 1993, followed by rapidly growing but unstable inflows since the end of apartheid. The success of the neo-liberal macro-economic strategy embodied in GEAR rests heavily on the assumption that large inflows of FDI will be forthcoming, but the country's limited ability to attract foreign capital has been the most disappointing feature of the economy since 1994. Investors' caution concerning the ANC's intentions is reflected in the low FDI totals for 1994-6 (Table 16.1). Rapid growth in 1997 was reversed in 1998, when South Africa was affected by growing negative sentiment towards emerging markets. An encouraging recovery in the second half of 1999 was dramatically reversed in 2000, when FDI fell 52 per cent to its lowest level since 1996, in part reflecting fallout from the growing crisis in Zimbabwe.

In contrast to the rest of Africa, where investment flows went primarily into minerals, South Africa attracts investment into manufacturing as well. The important contribution to attracting FDI made by the restructuring of state assets is also reflected in Table 16.1. The partial privatisation of Telkom in 1997 yielded $1bn. from Malaysian and US sources, and the partial privatisation of South African Airways in 1999 a further $0.23bn. Further privatisations will probably encourage increasing FDI flows over the medium term, but they are unlikely to equal the Telkom deal. In addition to the sectors noted in the table, motor and motor components industries are consistent performers, strengthened by the country's export strategies that have fitted into global outsourcing processes. Both foreign and domestic investment in hotels and leisure has declined since a peak in the mid-1990s, reflecting the view that demand needs to pick up (see below) before further investment is needed, but government award of gaming licences has increased investment in this sub-sector (Heese, 2000).

Table 16.1 Foreign Direct Investment in South Africa, 1994-2000, in US$m.

Year	FDI	Major investors (country and sector)
1994	1,723	-
1995	1,573	Coca Cola (USA – beverages)
1996	2,025	Petronas (Malaysia – oil)
1997	3,417	SBC Communication/Telkom Malaysia (USA/Malaysia - telecommunications); Dow Chemicals (USA - chemicals)
1998	3,251	Petronas (Malaysia - oil), Lafarge (France - construction), Placer Dome Inc. (Canada - mining), Lonrho Plc (UK - mining)
1999	4,092	Swissair (Switzerland - airways), A P Moller (Denmark - transport)
2000	1,968	-

Source: BusinessMap on-line database and Heese (2000).

US investments now outstrip those of any other single country, reaching R25bn. in 2001, of which R1bn. was spent on corporate social investments (*Business Day*, 2001). Malaysia is second, followed by the UK, Switzerland, Germany, Japan, France, Italy, Canada and Denmark. The list of smaller investors, including Kuwait, Indonesia, India, Dubai, Singapore and Thailand, suggests avenues for future relationships and confirms the potential importance of the Indian Ocean Rim.

Overall, however, FDI remains a disappointingly small proportion of total investment. According to the Reserve Bank, which makes a strict distinction between FDI and portfolio investment, South African FDI abroad actually exceeded that received from foreign investors by R15bn. in the period 1991-8. By the end of 1998 total FDI abroad amounted to

R170bn., compared with only R92bn. of FDI in South Africa. Europe accounts for 71 per cent of FDI in South Africa and a massive 89 per cent of South African investment overseas; the UK alone is responsible for 39 per cent and 34 per cent respectively (Lemon, 2000b).

Portfolio investment in South Africa is, in contrast to FDI, double that of South Africans abroad: for Europe this ratio increases to 2.75. But such investment does little to increase production and growth, create jobs and introduce new management techniques, skills and technology. Portfolio investment is also easily withdrawn, and capital flight, both legal and illegal, is a serious problem. According to official figures R17.4bn. left the country between the inception of an investment allowance in July 1997 and the end of 2000 (Manuel, 2001). Nearly 74,000 people took advantage of the opportunity to move R750,000 out on a one-off basis, and the government did not, as had been expected, raise the limit to R1m. in the 2001-2 budget. The amount leaving through other routes is almost certainly much larger, as individuals and businesses find means to ship wealth abroad, and because of the establishment in South Africa of international crime syndicates from China, Nigeria, Morocco, Russia and Colombia to take advantage of South Africa's currently lax regulations. New legislation intended to restrict money laundering is expected to come into operation in 2002.

Foreign Aid to South Africa and its Neighbours

The shortcomings of official development assistance are well recognised: at worst 'Northern aid has bandaged up what Northern aggression has savaged' (Sogge, 1998, p.55). Whilst in the short term aid may help to preserve institutions and stabilise political regimes, its political effects may be destabilising in the longer term, especially when accompanied by the conditionality imposed by international financial institutions. Quantitatively, however, there is no gainsaying the importance of aid to sub-Saharan Africa as a whole, where it was nine times as great as private foreign direct investment in 1997 (ibid., p. 51). In southern Africa, whilst earlier hopes of a 'Marshall aid' plan for the region after the demise of apartheid failed to materialise, aid remains important in relation to government spending (Table 16.2). It exceeds government spending from domestic sources in Mozambique, where it has been critical to the country's reconstruction since the coming of peace and holding of free elections in 1994, and in Malawi. It has also formed an important component of available resources in Zambia, Lesotho, Zimbabwe,

Namibia, Swaziland and Angola, but much less so in South Africa and Botswana.

Table 16.2 Foreign Aid in Southern Africa as a Percentage of Government Spending from Domestic Sources, 1996

Angola	21	Mozambique	336	Swaziland	29
Botswana	8	Namibia	47	Zambia	86
Lesotho	76	South Africa	6	Zimbabwe	34
Malawi	263				

Source: Sogge (1998).

Aid levels are influenced by many factors other than need, with the result that they show little correlation with poverty. In per capita terms all South Africa's neighbours received levels of aid well above the sub-Saharan average in 1995 (Table 16.3). Botswana was rewarded by aid donors for the stability of its democracy and its macro-economic policies, notwithstanding its relative wealth. Namibia, with by far the largest per capita aid level, was still in its first decade of independence. Both these states, along with Lesotho and Swaziland, have small populations, an attribute generally correlated with higher levels of per capita aid (Lemon, 1995). In Lesotho, Zimbabwe and perhaps in Swaziland too the effects of political instability are reflected in the decline of aid in the late-1990s.

Table 16.3 Foreign Aid Per Capita (US$) to South Africa and its Neighbours, 1995-1999

Country	1995	1998	1999
South Africa	9.9	12.4	12.8
Botswana	61.7	68.1	38.3
Lesotho	58.9	32.2	14.8
Mozambique	67.3	61.3	6.8
Namibia	124.2	108.3	104.4
Swaziland	61.6	30.7	28.4
Zimbabwe	44.6	24.0	20.5
Sub-Saharan Africa	32.6	22.6	19.5

Source: The World Bank.
(http://devdata.worldbank.org/external/dgprofile.asp?rmdk=82692&w=0&L=E)

In South Africa, not surprisingly, aid levels have increased massively since the end of apartheid. In pledging aid, donors have several goals: to identify themselves with the policy of racial reconciliation; to help stabilise a new multi-party democracy; to contribute to socio-economic redistribution in the aftermath of apartheid; and to endorse economic reforms that could serve as an example for the rest of Africa (Bratton and Landsberg, 1999). Western aid to South Africa should also be seen in the context of donors seeking strong allies and reliable trading partners in developing areas.

South Africa's thirty or so multilateral and bilateral donors probably pledged about $6bn. in development aid for the period 1994-99, of which loans represented about one-third (ibid.). Although tiny in relation to total government spending, they represented more than twice the domestic funds assigned to the Reconstruction and Development Programme in the mid-1990s (Bratton and Landsberg, 1998). Prior to 1994 aid flowed mainly into urban and peri-urban areas where civic organisations and self-help NGOs had been most active, but NGOs have seen their level and share of development aid fall since 1994 as aid has increasingly been diverted to official channels. Donors have been active in the following six sectors, in order of importance:

- education (primary, secondary and tertiary);
- democracy and governance (including human rights and public administration);
- agriculture (including water development and natural resource management);
- business development (especially SMMEs);
- health (especially primary and preventative care);
- housing (including infrastructure).

(Bratton and Landsberg, 1999)

The Department of Finance has encouraged donors to concentrate on areas such as the Eastern Cape, Mpumalanga and the Northern Province which are experiencing serious resource and capacity constraints.

The EU has consistently been South Africa's largest donor, although it has much lower than average rates of converting pledges to commitments, and commitments to actual disbursements (ibid.). Other donors, including the USA and Japan, have a much better record in both respects. Gaps between aid promised and aid delivered are partially a reflection of the donors' own procedures, especially in the case of the EU. For other donors, including small or medium-scale donors who can move money rapidly, the major frustrations arise from South Africa's own procedures, conflicts between donors and government, the gradual demobilisation of civil society, and above all weakness in the capacity of state institutions. Other problems arise from the nature of the assistance offered: some donors have offered large programmes that are too inflexibly designed or too slow-moving to alleviate pressing social needs in the short term (Bratton and Landsberg, 1998). Pretoria has established binational commissions with several of its key aid partners as a forum for resolving such policy differences.

The European Union

Since 1994 the EU has become South Africa's largest trading partner and foreign investor as well as aid donor. In 2000, over 40 per cent of South Africa's exports went to the EU and finance from the Union accounted for 70 per cent of South Africa's total foreign direct investments (European Union in South Africa, 2001). The EU's development programme in South Africa, the European Programme for Reconstruction and Development (EPRD), has an average annual budget of 130m. euro (approximately R1bn.), making it the largest single development programme in South Africa and one of the biggest implemented by the EU in any part of the world. Clearly, the European connection is of vital importance to the future economic prosperity of South Africa.

Following South Africa's first democratic election in 1994, the EU agreed to promote a 'package of immediate measures' designed to support South Africa's transition to democracy and its Reconstruction and Development Programme (RDP). By July 1994, all European sanctions had been removed. However, the ending of sanctions did not reduce significantly South Africa's relatively disadvantageous position in the hierarchy of access granted to the EU market. Since its creation in 1957, the EU has developed a complex network of preferential trading agreements designed to 'favour' selected states (Figure 16.1). Somewhat bizarrely, given its name, the EU's most favoured nation (MFN) status is at the very bottom of the privilege pyramid, alongside two levels of the generalised system of preferences (GSP). Up to 1994, South African access to the EU market was, with the exception of 3.5 per cent of products affected by European sanctions, governed by its MFN status. South Africa was therefore accorded market access on terms similar to the world's most developed nations; the USA, Canada and Japan. In 1995 Steven and Kennan (1995, p. 13) noted:

South Africa's current negotiations with the EU on market access are not necessarily concerned with being given 'preferential treatment'... On the contrary, they are in the first instance concerned with removing discrimination against South Africa and in favour of states that are, in some cases, richer and highly competitive.

Figure 16.1 The European Union's Pyramid of Trading Preferences

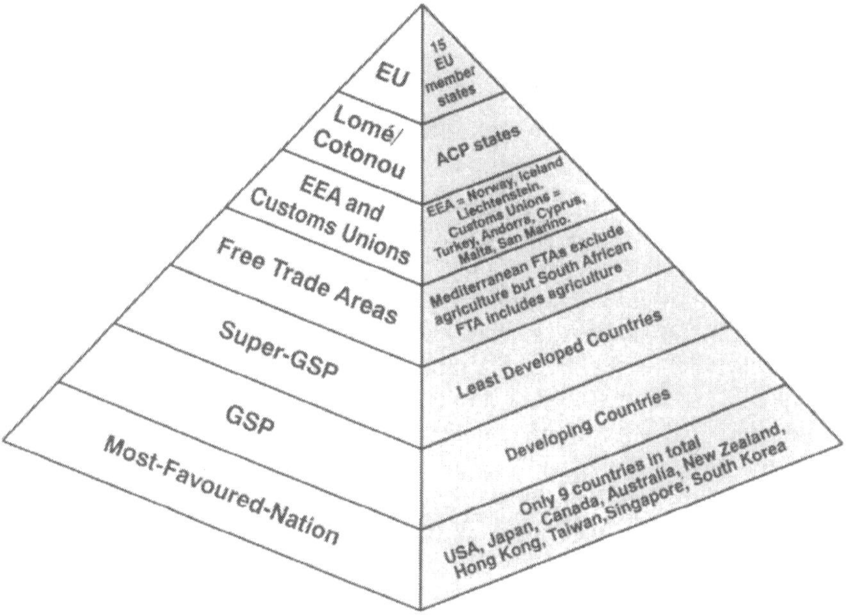

South Africa and the Lomé Convention

The Lomé Convention was an agreement linking the 15 member states of the EU to the 71-member African, Caribbean and Pacific group of states (ACP states). The countries belonging to the ACP under Lomé, 48 in Africa, 15 in the Caribbean and 8 in the Pacific, were the ex-colonies of the UK, France, Portugal, Belgium and the Netherlands. Under Lomé, all industrial products had a tariff exemption and most agricultural exports benefited from quota regimes or commodity protocols. A distinguishing feature of the Lomé trading regime was its non-reciprocity, allowing the ACP to erect and maintain tariff barriers against EU goods and services whilst being given duty free access to the Union market (Gibb, 2000).

In November 1994 South Africa requested the establishment of 'the closest possible relationship to the Lomé convention' and formally applied for membership (Holland, 1995a, 1995b) on the grounds that:

- It would provide South Africa with the same level of market access accorded to the thirteen other states of the Southern African Development Community (SADC) and would, as a result, facilitate intra-regional trade and investment.
- It would ease any diplomatic tensions between South Africa and the other countries of southern Africa by including these states, all signatories to Lomé, in the EU-South Africa negotiations.
- Lomé would offer South Africa the greatest trade preferences and, in so doing, support the RDP's aims to promote sustainable and equitable growth.
- Lomé IV was in any case due to expire in 2000 during which time South Africa's economy was expected to have experienced a substantial transition away from protectionism.

South Africa's Lomé application was supported by the European Parliament, the ACP countries, SADC and the Southern African Customs Union (SACU) (Graumans, 1997). The latter three perceived the benefits of South African membership in terms of enhanced bargaining power in the negotiations with the EU to replace Lomé. According to Kibble et al. (1995) it also reflected the lack of direct economic competition between the ACP states and South Africa. While many NGOs and parts of the academic community were supportive of South Africa's bid for Lomé status, others were strongly dismissive arguing that Lomé might even damage South Africa's long-term economic prospects by promoting protectionism and acting as a disincentive to FDI (Holland, 1995b).

In November 1994, the same month as South Africa's application, the EU rejected Pretoria's request to access the *full* provisions of the Lomé Convention. According to European documentation, the decision to exclude South Africa from full Lomé membership was based, in part, on the dual nature of the South African economy which more closely resembles that of a 'developed' than a 'developing' country (European Commission, 1997, p. 6). The EU argued that as the non-reciprocal trade provisions of the Lomé system were specifically designed to assist the development of some the world's poorest states, South African membership was inappropriate. The EU advanced four key arguments in support of its policy of rejection:

- South African membership of Lomé would erode the benefits accruing to the existing ACP states. In a Lomé context, South Africa has a relatively diversified and extremely competitive economy significantly larger than any ACP country. In 1995, South Africa's total exports to the EU were equivalent to more than one-third of the exports of all 70 ACP countries combined.

- Allowing South Africa full access to the provisions of Lomé would threaten EU producers, in both agriculture and trade, and, in so doing, undermine support for the continued non-reciprocal trade preferences of Lomé.
- South Africa's accession to Lomé would, according to the Commission, be challenged by the WTO as promoting further discrimination against other non-Lomé developing countries as well as countries in transition.
- Finally, the Commission argued that Lomé might have presented South Africa with an opportunity to slow down the pace of its liberalisation programme and, as a consequence, promote protectionism and act as a serious disincentive for potential foreign investors.

With South Africa's Lomé application rejected, the Council of Ministers, the EU's supreme decision-making body, approved a mandate for the European Commission to negotiate a framework for long-term relations with South Africa. The 1996 mandate called for a framework built around two pillars: a protocol offering South Africa *qualified* membership of the Lomé Convention and a bilateral agreement for trade and co-operation.

South Africa's qualified membership of Lomé was approved at a joint meeting of the ACP/EU Council of Ministers in Luxembourg on 24 April, 1997 (Laidler, 1999). On 23 June 2000, a new partnership agreement between the EU and ACP was signed in Cotonou, the capital of the west African state of Benin. Cotonou will, over a transitional period lasting eight years, replace the trading arrangements established by the Lomé Convention. However South Africa, although being a member of Cotonou, does not have access to its trading preferences. South African benefits include, *inter alia*, technical, cultural and social co-operation, regional co-operation, participation in Cotonou tenders, regional projects, refugee assistance and cumulation of origin rules.

The EU's mandate of 1996 also called for a comprehensive bilateral trade and co-operation agreement covering all those subjects not dealt with in the context of South Africa's qualified membership of Lomé, the most important component of which concerned the establishment of a free trade area (FTA).

The Trade Development and Co-operation Agreement (TDCA)

The TDCA was signed by the EU's Council of Ministers and ratified by the South African Parliament in October 1999, and operationalised on 1 January 2000. The most ambitious provisions of the TDCA are those concerning the establishment of a FTA. However the TDCA is concerned also with co-operation in the areas of trade-related issues, political dialogue, development co-operation, social and cultural co-operation, financial assistance and economic co-operation. Three separate sectoral agreements were also attached to the TDCA, covering science and technology, wines and spirits, and fisheries (Figure 16.2).

Figure 16.2 EU-SA Trade, Development and Co-operation Agreement

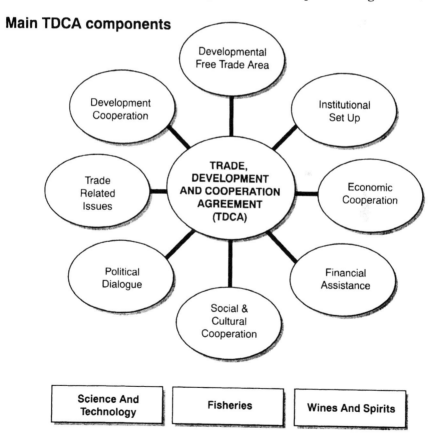

Although some co-operation agreements were resolved swiftly, such as the Science and Technology Agreement in December 1996, TDCA negotiations were characterised more by acrimony than harmony. The FTA negotiations were particularly difficult, extending over four years and involving 24 rounds of official face-to-face discussions. The main details of the trade pact were agreed by the EU at its Berlin summit in March 1999 and signed by the Council of Ministers on 11 October 1999. However in 2002 negotiations continue over two co-operation agreements linked to, but outside of, the TDCA, namely wines and spirits, and fisheries.

The FTA, although described as being a reciprocal free trade agreement, will be differential in coverage and asymmetrical in timing. EU literature (European Union in South Africa, 2001) refers to the FTA as being 'developmental' in character, allowing South African exports better access over a shorter period of time than European exports to South Africa. Specifically, the EU will liberalise 95 per cent of its South African imports within ten years while the respective figures for South Africa are 86 per cent in twelve years (Figure 16.3). Within these transitional periods the bulk of liberalisation of industrial products on the EU side will take place in the first three years, whereas on the South African side tariff reduction for most industrial products will only start after six years (Figure 16.4). As observed by the European Union in South Africa (2001, p. 8):

> Asymmetry and differentiation: these twin concepts reflect the developmental approach of the trade provisions. In recognition of South Africa's economic restructuring efforts currently underway, the EU will open up its market faster and more extensively for South African products than it will ask South Africa to do for EU products.

Both the EU and South Africa have excluded certain products from the FTA in order to protect economically vulnerable or sensitive sectors. In addition, certain products are subject to only partial liberalisation. On the South African side the main sensitivities lie in the industrial sector while the EU has made exemptions mainly on agricultural products (Table 16.4).

The Developmental Component

An explicit objective of the TDCA is to support economic development in South Africa and, indirectly, southern Africa. Although the agreement is innovative in several respects, it stands out as being one of the EU's first reciprocal trade agreements linking the Union to a developing/transitional country. As such, the TDCA symbolises an important shift in the EU's development strategy away from interventionism and non-reciprocity towards neo-liberal free trade policies. The development component of the FTA therefore rests with the asymmetrical timing and coverage.

Figure 16.3   EU-SA Free Trade Agreement: Asymmetrical and Differentiated

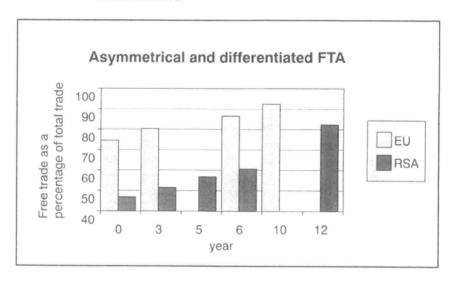

Figure 16.4   EU-SA Free Trade in Industrial Products

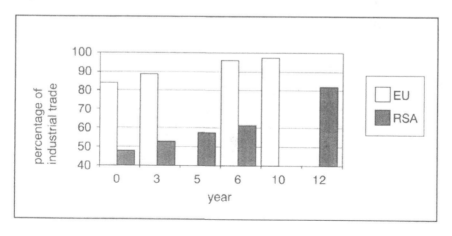

Table 16.4 Sensitive Produce Excluded from the EU-SADC Free Trade Area

EU exemptions		SA exemptions
Meat	Cereals	Footwear and clothing
Vegetables	Flowers	Motor vehicles and
Fruit	Sugar	components
Cut flowers	Fruit juices	Textiles
Dairy products	Starches	
Maize	Wine and vermouth	

Whilst the proposed trade liberalisation package for industrial products may at first sight appear generous, Goodison (1999) points out that because of the already low tariff structure in the EU, together with the fact that the majority of South African exports already enter the EU market duty free, the adjustment costs of implementing such an agreement will be far greater for South Africa. Goodison (1999) calculates that to achieve 90 per cent duty free access the EU has to eliminate duties on approximately 7 per cent of its South African imports beyond what it is already committed to do in terms of the GATT Uruguay Round. The equivalent figure for South Africa is 36 per cent.

The agricultural negotiations proved to be the most acrimonious and sensitive. In order that the FTA comply with the principles of the CAP and not affect the practice of EU preference, the original 1996 mandate proposed to eliminate duties on only 60 per cent of South Africa's current agricultural exports. The Union produced a list of agricultural produce to be excluded from the FTA, the total value of which equalled 39 per cent of the EU's agricultural imports from South Africa. Furthermore, the EU produced a list of sensitive agricultural produce for which little trade has been recorded and no liberalisation was envisaged. In return the EU asked South Africa to liberalise its agricultural imports in two stages and on 95 per cent of all EU agricultural exports. Bluntly stated, the EU expected South Africa to eliminate duties on nearly all EU agricultural exports while retaining European protection against nearly 40 per cent of South African exports. Whilst the EU emphasised that the proposed agricultural exclusions represented less than 4 per cent of total EU imports from South Africa, Pretoria saw agriculture as an area in which it would have a competitive advantage in the EU market. The final agreement on agriculture (Figure 16.5) committed the EU to liberalising 61.4 per cent of existing South African agricultural imports with an additional 13 per cent being subject to partial liberalisation.

Figure 16.5 EU-SA Free Trade in Agricultural Produce

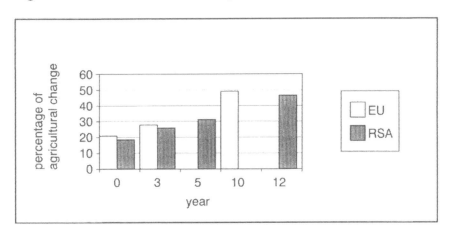

EU-South Africa Trade Data, 2000-2001

EU imports from South Africa grew substantially in the first half of 2001, by 32 per cent compared with the same period in 2000, whereas EU imports from the rest of the world grew by only 10 per cent in the same period. Furthermore these data are calculated in euro and do not, therefore, reflect the weaker value of the rand exaggerating South African exports. In monetary terms, the 2001 increase is worth almost 2bn. euro and brings total EU imports from South Africa to above 8bn. euro. The trade surplus in favour of South Africa broadened further in the first half of 2001 to 1.9bn. euro compared with 0.7bn. euro in the same period of 2000.

Since 2000, South Africa's rapid export growth has been spread over a wide range of sectors including traditional primary exports, particularly platinum, diamonds and coal, as well as agricultural products and machinery. Agricultural products showed very positive development, with tea and coffee emerging as a new and strong export market doubling in value to 23m. euro. Machinery now constitutes 10 per cent of South African exports to the EU, with motor vehicles contributing another 4 per cent. EU exports to South Africa also experienced positive growth compared with exports to the rest of the world: 14 per cent in the first six months of 2001 compared with the same period in 2000. Motor vehicles performed most strongly, registering a 43 per cent increase.

Conclusion

Together chapters 15 and 16 reveal a series of unequal trading relationships. At the regional level South Africa dominates completely the political economy of southern Africa. The commodity composition of southern African trade reflects a classic developed-developing country relationship, with South Africa exporting manufactured goods and importing primary and unprocessed raw materials. South Africa, partly as a consequence of this commodity composition, maintains a considerable trade surplus with southern Africa. An unequal trading relationship also exists between the EU and South Africa. First, the South Africa-EU trading regime is far more important to South Africa than it is to Europe. For South Africa the creation and maintenance of a favourable and stable trading relationship with the EU is of critical importance to the economic well-being of the country. For the EU it is of secondary importance. In addition, and similar to the South Africa-southern Africa trading relationship, the commodity composition of South Africa's trade with the EU is unbalanced, with the EU exporting manufactured goods and importing mainly primary products and semi-processed materials. This pattern of inequality and dominance lends weight to the EU's classification of South Africa as a 'country in transition', being neither developed nor developing in status. Using a world-systems framework South Africa is a classic example of a semi-peripheral country, bridging the gap between the core and peripheral worlds and representing the interests of capital in the periphery. The neo-liberal trade policy adopted by the ANC since 1994 may or may not generate the growth needed to push South Africa towards developed country status. Only time and geography will tell.

References

Ahwireng-Obeng, F. (1997), 'A Sceptical View of South Africa within the IOR-ARC', *South African Journal of International Affairs*, vol. 5, pp. 97-109.

Bratton, M. and Landsberg, C. (1998), 'Trends in Aid to South Africa', *Indicator SA*, vol. 15 (4), pp. 59-67.

Breytenbach, W. J. (1999), 'Indian Ocean Rim: Assessing the Prospects for Co-operation and Integration', *Development Southern Africa*, vol. 16, pp. 89-105.

Business Day (2000a), 'South Africa Making Comeback as Exporter of Automotive Parts', 19 September.

Business Day (2000b), 'United States Visualises Major Trade Pact with Africa', 14 November.

Business Day (2000c), 'Free Trade Agreement between South Africa and Brazil could Backfire', 7 August.

Business Day (2000d), 'South Africa Considering Accords with Nigeria and Libya', 24 July.

Business Day (2001), 'Recent US Investments Reach R25bn', 25 May.

European Commission (1997), *The European Union and South Africa: Building a Framework for Long Term Cooperation*, Task Force South Africa in DGVIII, Brussels.

European Union in South Africa (2001) *EU-SA Trade and Economic Cooperation*, www.eusa.org.za (accessed 12 May 2001).

Fine, B. and Rustomjee, Z. (1996), *The Political Economy of South Africa: from Minerals-Energy Complex to Industrialisation*, Westview, London.

Gibb, R. A. (2000), 'Post-Lomé: The European Union and the South', *Third Word Quarterly*, vol. 21, pp. 437-81.

Goodison, P. (1999), *Marginalisation or Integration? Implications for South Africa's Partners of the South Africa-European Union Trade Deal*, Occasional Paper no. 22, Foundation for Global Dialogue, Johannesburg.

Graumans, A. (1997), *Redefining Relations between South Africa and the European Union*, Occasional Paper no. 10, Foundation for Global Dialogue, Johannesburg.

Heese, K. (2000), 'Foreign Direct Investment in South Africa (1994-9) – Confronting Globalisation', *Development Southern Africa*, vol. 17, pp. 389-400.

Hirsch, A. and Hanival, S. (1998), 'Industrial Restructuring in South Africa: the Perspective from Government', paper presented at the 1998 Annual Forum, Trade and Industry Policy Secretariat, Gleenburn Lodge, Muldersdrift, South Africa.

Holden, M.G. and Isemonger, A.G. (1999), 'A Review of Trade Trends: South Africa and the Indian Ocean Rim', *Development Southern Africa*, vol. 16, pp. 69-88.

Holland, M. (1995a), 'South Africa, SADC and the European Union: Matching Bilateral with Regional Policies', *The Journal of Modern African Studies*, vol. 33, pp. 263-83.

Holland, M. (1995b), *European Union Common Foreign Policy: from EPC to CFSP Joint Action and South Africa*, Macmillan, London.

ITEDD (2000), *Multilateral Strategy 2000/01*, International Trade and Development Division, Department of Trade and Industry, Pretoria.

Kibble, S., Goodison, P. and Tsie, B. (1995), 'The Uneasy Triangle - South Africa, Southern Africa and Europe in the Post-apartheid Era', *International Relations*, vol. 14, pp. 41-61.

Laidler, M. (1999), 'The European Union-South Africa Trade, Development and Cooperation Agreement' in *South African Business and the European Union in the Context of the New Trade and Development Agreement*, Seminar Report, Konrad-Adenauer-Stiftung, Johannesburg, pp. 13-17.

Lemon, A. (1995), 'Small States in the Modern World', *Geography Review*, vol. 49 (2), 24-29.

Lemon, A. (2000a), 'New Directions in a Changing Global Environment: Foreign Policy in Post-apartheid South Africa', *South African Geographical* Journal, vol. 82, pp. 30-39.

Lemon, A. (2000b), 'South Africa's Relations with the EU since the End of Apartheid', *Tijdschrift voor Economische en Sociale Geografie*, vol. 91, pp. 451-57.

Manuel, Trevor, Finance Minister (2001), Statement to the South African Parliament, reported in *Southscan*, vol. 16 (7), p. 1.

Moodliar, S. (1997), 'The IOR-ARC: how it will Benefit SA', *Global Dialogue*, vol. 2 (1), pp. 14-15.

Roberts S. (2000), 'Understanding the Effects of Trade Policy Reform: the case of South Africa', *South African Journal of Economics*, vol. 68, pp. 607-38.

Sogge, D. (1998), 'Misgivings: Aid to Africa', *Indicator SA*, vol. 15 (4), pp. 51-8.

Southscan (2000), 'Mbeki seeks to Revitalise Renaissance on Tour', vol. 15 (20), 6 October, 154-55.

Stevens, C. and Kennan, J. (1995), *Trade between South Africa and Europe: Future Prospects and Policy Choices*, Working Paper no. 26, Institute of Development Studies, Sussex.

Tsikata, Y. (1999), *Liberalisation and Trade Performance in South Africa*, Informal Discussion Papers on the South African Economy no. 13, World Bank, Washington DC.

Van Seventer, D. E. (2000), 'Inter-industry Revealed Comparative Advantage for South Africa', *Focus on Data*, December 2000, Trade and Industrial Policy Secretariat, University of the Witwatersrand, Johannesburg.

Index

Printed and bound by CPI Group (UK) Ltd, Croydon, CR0 4YY

21/10/2024

01777082-0007